Anonymus

Vorlesungen über anorganische Chemie

Anonymus

Vorlesungen über anorganische Chemie

ISBN/EAN: 9783742818430

Manufactured in Europe, USA, Canada, Australia, Japa

Cover: Foto ©Angelika Wolter / pixelio.de

Manufactured and distributed by brebook publishing software (www.brebook.com)

Anonymus

Vorlesungen über anorganische Chemie

Lecture 68th

Charcoal. Stone & Brown Coals. &c are all impure modifications of Organic Carbon

To prepare Charcoal we ~~burn~~ heat, the mass of wood piled in covered places, & stacks- that is with the presence of an insufficient amount of Oxygen - the consequence is, that the other constituents H & O &c. of the wood are consumed & go off as HO-&c. while - the Carbon is left behind - each particle in the place it occupied in the living wood: so that the form of the fibre & the direction of vessels is perfectly retained - & from a coal we are able to decide the tree species from which it came. Wood fibre

Adaptation of diff. Coals for Differ. uses

There are some Charcoals that cannot be used in the laboratory - others can - for ex: - Oak - charc. cannot be used in Chemistry - because it decrepitates - & the danger is present that our vessels may become impurified with the pieces of our fuel. Pine &c &c. are much specifically lighter & much better adapted to our uses. If animal substances (bones) be treated in a similar manner as the wood, we obtain a still more impure form of Carbon. - like wood - we here still retain the form of the bones; this is called Animal Charcoal, & is of the most extended importance in manufacture.

Animal Charcoal

When the products of the Lamp-
Combus. of subst. rich in Carbon, Black.
which burned in insufficient draft-
are lead into cool chambers - a
great amt of unconsumed Carbon
settles upon the walls, in the form
of an impalpable powder -
this is almost pure C.
& finds its way into the Combus-
Markets under the name tibility
of Lampblack. Thus col- Danger
lected on the walls it is very dan- of Pack-
gerous to pack it - for fear of ing.
spontaneous Combustion.
This Carbon is almost totally
free from impurity. This form
(the amorph. generally) is very indifferent to the action of reagents at
ordinary temps, can be left ly-
ing for Years - exposed to the action
of heat, atmosp. & moist. without change.

635.

We make use of this property—
when we ~~burn~~ char the bottoms of
logs & timber which must be
placed in the ground.

One singular property—of which
Decolor- the most extensive practical
ization use is made—is the power of them
of Organ- amorphs. C— & partic—Animal C.
ic Matters of seizing upon—& fixing upon
Ex itself—Organic Coloring matters—
(Why? has not been Explained)
Used in Sugar refining—
Perform Exper—of decolorizing
Red Wine. Indigo—&c. This is
not brought about by Chem. Union.
By thus filtering a solution re-
peatedly, containing, a coloring
matter—a gum, & sugar—
We will find that first the
Coloring matter will be depos-
ited upon the coal—then

by separating the gum, & lastly the sugar.
This substance has also the power of absorbing gases in a very high degree. With freshly glowed Coal & NH_3 the absorption can be easily shown. Upon this property depends the antiseptic virtues of Carbon, viz: its use in preserving animal matters - meats &c from decay - Uses viz - When packed in C. Cupons (sea-voyages partic). For this purpose freshly glowed Wood Coal is used finely powdered. - again - it is used in the same way as a disinfectant owing to its power of absorbing - gases. Hence the danger of allowing larger quantities of fine C. together - combustion &c.

Eg

Uses of C.

in techni

637

We have, now, considered the properties of the various modifications of C. + we have now to show that the product of

CO

the combustion of Diamond + of Graphite + of Charcoal is one + the same, thus giving an ocular + Chemical demonstration of the fact that they are one + the same substance.

Combustion of Charcoal, Graphite + Diamond.

The latter inflames the easiest + introduced into Oxygen burns (glows) on with a tolerable light.

The Prod. of Combustion CO_2

Graphite must be intensely heated in the Oxy-Hydrogen blowpipe before it will combine + then only with difficulty.

Diamond requires the same blowpipe heating, but afterward (as proved by tests) combines readily. CO_2 is formed.

Carbon + Oxygen.

CO_2 = Carbonic acid.
C_2O_3,HO = Oxalic acid.
CO = Carbonic Oxide.

CO_2 -

The most important of all compds - except O - for the ~~Orgy~~ Organic World. It combines the organic world with the Inorganic. It is produced in many parts of the world by springs - & at times - ~~by the~~ as gas, it issues in immense quantities, from fissures &c &c in the solid Crust of the Earth. So in the Dog's Grotto near Naples - where the gas issues from fissures in the rocks - & forms a stratum of several feet in thickness upon the floor.

Illuminates.

The Dog's Grotto

By Volcanic Action

Volcanic action is always attended by the abundant evolution + manufacture of CO_2 (by the action of acid gases + heat upon $CaOCO_2$ &c) - a fact attested by repeated observation.

Again, it is found free in the atmosphere - forming 1/3000 of its volume -

Again - as a constituent of Limestones - Dolomites &c - which form in all parts of the world immense mountain masses.

Animals, Plants + CO_2

The most important connection exists between this gas + the existing (+ pre-existing) organic world. The animals require for their support, Oxygen. + as refuse product they expire CO_2 - + HO; this

CO_2 which, by its accumulation in numberless ways, would soon produce death — is the very product which the plants require for their sustenance, (+ they give off O) + thus by this mutual interchange of O + CO_2 — the maintenance + Equilib. is maint'd.

Can be manufactured by the direct union of the substances — O + C.

Manufacture.

By this Combustion there is no change in the volume for upon Cooling the volume is the same

Eq.

Hence, the composition by volume is the following.

Comp. by Volume

2 vols. O = 2.2112
1 vol. C = 0.8468
2 vols CO_2 = 3.0580
1 vol. CO_2 = 1.529.

The sp. gr. of C. is obtained indirectly by the foll. means.

641

The Sp. Gr. of C. vapor.

773cc. CO_2 = 1.52909 grms.
773cc O = 1.1056 "
C in 1.529 CO_2 = 0.4234

The equiv. vol. of O = 1. ergo :- as the formula is CO_2 - with 2 × 1.1056 O_2 = 2.2112 gr. we must have 2 × 0.4234 C = 0.8468.
1 Eq. CO_2 = 2 vols = $\frac{3.0586}{2}$ = 1.5298.

Manufacture from $CaOCO_2$

This method of manufacture is, however never, attempted in the laboratory - on acct. of its inconvenience - The method we adopt is to treat a Carbonate (generally $CaOCO_2$ with a dilute acid - HCl for example, whereby the following takes place

$CaOCO_2$ } CaCl + 2HO
HCl.HO } CO_2

Conc. SO_3 the best Acid to use

With dilute SO_3 - a surface strata of insoluble $CaOSO_3$ ad. forms about the $CaOCO_2$ - + hinder the gas generation. but with Conc. SO_3 - it is best. Why? next lecture

Lecture 69th
Carbonic Acid. cont'd.

With HCl HO – there is always more or less HCl which goes over, & our gas is somewhat purified. With SO_3 HO, an insoluble CaO SO_3 is formed which hinders the evolution of the gas. When, however, Conc. With SO_3 SO_3 + little HO is used, the solid & little HO CaOCO₂ by constant decrepitation (from the generated heat) is always brought into contact with fresh SO_3 – this is the best method.

At times we frequently need to carry on analytical operations in vessels entirely free from atmosphere – for fear of Oxidation, this is accomplished by means of the ordinary CO_2 apparatus use in the laboratory – Explain

643.

Testing air for CO_2 — It is often necessary to test the air in deep wells - on acct of the danger of descending into them - a flask filled with + dipping into a vessel of Hg is used. when lowered it can be raised + filled with the air.

Properties — It is a colorless, transparent gas - which can be condensed into a fluid, at a temperature of -78°C. or by a pressure of from 30-50 atmospheres.

As a liquid, it is a colorless transparent fluid. which exposed to air evaporates rapidly enough to solidify the ~~gas~~ liquid.

The Aparatus described — The aparatus is the common condensing pump - The collector - however made of the best + strongest wrought Iron.

By allowing it to escape into a closed vessel – it condenses to a white snowlike solid, which constantly evaporates – Brought upon the skin – it brings about the illustration of the 'Spheroidal state' the hand acting as the heater vessel ordinarily does. When pressed down upon the skin, however – it brings about a severe burn. The constant temperature is $-78°C$. Mixed with a substance with which it forms a liquid compound a more intense cold than this can be brought about ($-107°C$) with Sulphuric Ether) Hg can thus be frozen to a solid, & is in color like silver, in malleability like Pb.

Solid CO_2.

Brought on the Hand.

Spheroidal State

Freezing Hg. Exp.

§45

Liquid CO_2 — The liquid is perfectly colorless - oily - swimming upon H_2O

Sp. Gr. — & having a specific grav of 0.9989 at - 10.8°C. It expands enormously - even more so than the gas.

Carbonic Acid Gas.

Is much heavier than air - by bringing in it a lighted

Ex — taper, it is ~~to~~ extinguished. By reverting it & bringing the taper in it it burns.

It can be poured from one flask to another, Perform the Experiment of changing from one flask to another, air &

Ex — CO_2. By bringing a bladder of air into a vessel of CO_2, it will be supported upon it like a balloon in air. It can be weighed &c.

With moistened litmus paper the gas gives a slight acid reaction. It forms a great number of salts — well defined.

It ~~so~~ does not ordinarily support the combustion, of combustible substances. Ex. Papers are extinguished. Burning Only those substances which Na in CO_2 possess the greatest affinity for O. will burn in CO_2 Ex. CO forms. Na - g. b. but it must be highly heated beforehand. H reduces CO_2

CO_2

$H + CO_2$ passed over a glowing tube of Porcellain acts to bring about a reduction to CO.

The absorption's coefficient of CO_2 is regular — + the amt. of CO_2 at the various temperatures is entirely dependent upon the law of the absorption. At 2 atmospheres twice as much gas as at one atmosphere.

Mineral Waters

When, therefore, we wish to prepare ourselves a large quantity of gas in HO. we do so, under a pressure of from 3 to 5 atmosph. So our ordinary mineral Water is formed.

It is found in Rain Water – it plays an important part in the changes, in the rocks of the earth's surface.

Action of CO_2 upon Rocks

There is always a continual filtering process going on, the acid waters permeating the rocks – & dissolving out soluble ingredients & carrying them with them – again to the surface & finally to the great reservoir – the sea. Again, the rocks are totally

Changed in their mineral character - Carbonates being formed - & silicates of Alkalies carried off. The immense beds of $CaO\,CO_2$ are being constantly worn away & their ingredients carried into solution - to the Ocean for in CO_2 HO - $CaO\,CO_2$ is soluble. & thus the supply for the organic uses of Corals & Mollusks kept up. By such & other complicated processes, carried on, not for thousands, but for millions of years - the ~~formation~~ of minerals - disintegration solution - & metamorphosis of the rocks have been accomplished - (see Bischoff - Geologie I Bd.)

Metamorphic Rocks

Ex

$CaO\,CO_2$ soluble in CO_2 excess

Lecture 76th

CO_2

Occurrence in Nature — forms many important salts. Many occur in nature - CaO CO_2 - MgO CO_2 &c.

They can very easily be detected ~~when~~ by moistening the suspected rock with a drop of strong acid (HCl &c) almost any other acid is strong enough to cause a violent effervescence of CO_2

The behaviour of the acid is in many cases that of a bi-basic

Type Bibasic —

$$\left.\begin{matrix} KO \\ HO \end{matrix}\right\} C_2O_4$$

acid, forming salts, after the type of the above salt.

Ex

absorption by KOHO &c. — By bringing a strong alkaline solution CaO, KO, NaO, &c. into contact with CO_2 - It absorbs the gas with great avidity, though the same

650

is not done so completely & rapidly as NH_4HO with H_2O. For the gas analyses we absorb the CO_2 by KO- & measure the amt. by the diminution of volume.

Gas Analysis

For the quantitative determination of CO_2 in mineral analyses- we make use of various apparatus- the one represented by the figure- is one of the best- It is weighed with the substance & CO_2 determined by the loss in weight.

Quantitative determinat. of CO_2.

Atmospheric air.

N	= 76.76000	Composition of the atmosphere by weight-
O	= 23.16995	
NH_3	= 0.00005	
CO_2	= 0.07000	
	100.000	

Besides these gaseous matters - there are minute quantities of solids always floating in it - not only dust, &c derived from the surface of the land, but the salt matters of the sea, derived from the dashing of fine spray etc - are always held in it, & carried over the land. Though very minute quantities are at any one time, yet when we take into ~~the~~ consideration lapse of time, the influence of these salts upon the fruitfulness of the soils - must be regarded as immense. These solid, mechanically mixed ingredients are made visible, by a ray of sunlight - when it enters a dark room.

Solid Matters in Air

Influence of these matters on Organic Life.

But more important for us are the other small ingredients in the air. This quantity of CO_2 & NH_3 in the air is proved by analysis to be constant. The method of determining these almost infinitely small quantities are the following.

NH_3 could only be present in infinitely small quantities. Absorption of — for — it is absorbed with so much avidity by H_2O — that NH_3 were it to issue from the depths in vast quantities — the amount in the atmosphere would not be increased. by H_2O

The exact quantity of NH_3 in the air has — (according to Bunsen) never been accurately determined.

653.

Analysis of air for NH_3 + of CO_2

The analysis of air - for its contents in CO_2 + NH_3 - is, as may be imagined - a difficult + important problem. The general principle is the following - The air is lead through a weighed or measured - quantity of HCl. of known strength - + lead through in such a way that we can tell with certainty how much air we have drawn through (best, by a graduated - Aspirator, or, by a gas meter. By a simple titration with a normal sol. of <u>NaHO</u> or by weighing - the amt of NH_3 absorbed can be accurately determined, for any given quantity of air.

The quantity of H_2O in the H_2O - a
air is the factor which variable
is exceedingly variable. Quantity
It is not - at times - con - in air.
stant for even an half hour.
For the CO_2 determ. we use
a similar meter - or aspirator -
& pass the dried air through
a snake formed tube of glass CO_2
filled with KOHO solution deter-
better - a weighed quantity or mination
(a measured one).
By weighing the appar-
atus before & after the
passing through the solu-
tion - the increase in weight
of the thing gives the absolute
weight of the CO_2 in the KO HO.
Or, if it be measured, a small measured
quantity can be introduced into
an ordinary CO_2 apparatus

655

& weighed with sufficient HCl. the loss in weight will give the amt of CO_2 in a known quantity of our KO solution

As the quantity of CO_2 in the air is of importance to the health &c of animals (& ourselves particularly) it is of interest to know if that quant. is a constant or a variable one. & analyses conducted most carefully show that – air in the most various places – from the north & ~~at~~ in the tropics – upon land & upon the sea – upon mountain-tops & in valleys – upon the plain & in the forest, all give for a result – an average & constant quantity of CO_2 – vz – about 3 parts by vol – in 10.000 parts of air.

CO_2 – about 3 parts in 10000 of air by vol.

Lecture 71st

An analysis of air — for its composition, relatively of O + N — is best carried out in the Eudiometer. (Perform an analysis — roughly over H_2O) Ex-p

The amt of O + of N in the atmosphere is a constant quantity, & not variable — as analyses from the most distant parts of the Earth & under all the possible variety of conditions have proven. Air from Tropics — & the poles — from many thousand feet above the earth — show that: —

Analysis of air from Diff. Places.

By vol Oxygen = 20.96 } is contained in
" " Nitrogen = 79.04
" " Atmosphere = 100.00

& that the variations only affect the second decimal figure.

657

It is a question of great interest for us to know whether there exist any causes which preserve the equilibrium of the atmosphere. One great destroyer of Oxygen — is the respiration of animals — all of whom take from the air the O + give it back again in the form of CO_2 + H_2O vapor — + this combination taking place in the body produces the phenomenon of animal heat. Another — is — the decomposition of organic bodies — which take oxygen from the air — to form CO_2 + c + c. and this process repeated for immense periods would of itself be sufficient to render the atmosphere, so poor in Oxygen that the organic world must change itself.

Causes which preserve the Equi-librium of the constituents of air.

again, the constant action of the Oxygen absorbed in HO— in permeating the rocks of the Earth's in carrying on oxidizing processes—would in time (geological) produce incalculable waste of this material. Lastly—through the agency of man—in bringing about for his comfort + support—artificial combustions, other vast quantities are lost + converted to CO_2— The great antagonistic to this consumption, + the one which profits by it,—+ derives its nourishment from this very consumption—is the 'Vegetable World.' Plants are the 'Great reducing agents' in Nature's laboratory.—for their growth they demand in—

Destroyer of Oxygen

Plants, the great reducing Agents of Nature

mense supplies of Carbon — & that supply they find in the CO_2 of the air. derived from animal respiration decomp. combustions & c&c: by aid of their leaves they absorb this gas, their organs assimilate the Carbon & the Oxygen is given back in its original state to the air, & thus the purity of the respirating medium maintained. A certain quantity of Oxygen, viz: that which held in meteoric Waters infiltrates into the earth, & carries on the great oxidizing processes with the rocks within its bosom, is lost, & the only restorers, of which we know are volcanoes — which by each eruption, exhaust great quantities of CO_2 — upon which the plants act as before mentioned.

The absolute loss of O. & its restoration by volcanoes.

note →

The question now arises. Is this complicated series of processes sufficient to ~~[illegible]~~ preserve the equilibrium of atmosph. Constitution,

A cursory view of the complicated conditions, seems to leave us no safe foothold from which to form a conclusion.

If we imagine the constituents of the atmosphere - present per se (separated) - the following numbers would illustrate their amt.

N. of average dens = 6432 m.
O " " " = 1657 "
HO " " " = 127 "
CO_2 " " " = 2.1 "

They suppose - these gases under the normal press. + temp — 0°C +, 760 mm. ~~at~~ upon the surface

661

A few harvests would be sufficient to remove from the Earth every trace of CO_2 from the air. We see from this, that the plants too, have only a small am't of CO_2 to depend upon, & vice versa, when Animals increase in great number - so also the CO_2 am't increases, hence: the most intimate connection exists between the two great parts of the organic World. An increase in the one - bringing with it the means for the correspondingincrease in the other. Thus acting mutualy as a check upon one another - & preserving a nicely poised & self regulating equilibrium in the inorganic world.

In our rivers & seas the organ- ic world is supported by ab- sorbed oxygen. The follow.g is the

Note. H₂O the process repeated.

Comp of Air in Water

$N = 65.10$

$O = 34.90$

Air in H_2O = 100.00

A considerable variation from the proportions of the atmos- phere. The abs. amt is of course upon temp. & pressure.

Lecture 72nd

The Ventilation of the H_2O of the Sea.

The streaming of the warm + cold waters of the ocean, from + towards the Equator - constantly tends to bring every particle of water into contact with fresh air - thus keeping it saturated with all it can absorb, + contributing to the necessities of the animals.

Temperat. of deep Waters 4°C'

If we make the same comparison of the gases in the Ocean as upon the Earth - we get a depth of gases of average density

note {
of N = 84.0 meters
of O = 45.0 "
of CO_2 = 3.6 "

It will be noticed that the amt of CO_2 is greater than upon the lands by one third.

It has been determined with certainty that all the light rays are necessary for the action of plants in decomposing CO_2 — a singular fact — for in the inorganic world — the chemical effect of light is confined to certain rays particularly the Blue + ultra violet — the rest being indifferent

All the light rays necessary for the action of Plants.

CO — Carbonic Oxide —

Manufactured by reduction of CO_2, viz — by leading CO_2 over glowing Coals — $CO_2 + C = 2CO$, By leading Oxygen over glowing coals — if the latter are present in excess — we can form the gas — if not — present in excess — CO_2 is formed — Neither of these ways are used in the laboratory —

Manufacture

665.

We manufacture it from the so-called Formic acid. (obtained from large ants upon cooking in HO) -

Manu facture This formic acid - treated with with an excess of $HOSO_3$, the following results

$$\left.\begin{matrix} HO, C_2HO_3 \\ HOSO_3 \end{matrix}\right\} \begin{matrix} CO \nearrow \\ CO \\ 3HOSO_3 \end{matrix}$$

Again - by Heating oxalic acid, or an Oxalate with

Ex / 11 SO_3HO - the same results - (~~same~~ the CO_2 is generated & must be absorbed by KOHO)

$$\left.\begin{matrix} HOC_2O_3 \\ HOSO_3 \end{matrix}\right\} \begin{matrix} CO_2 \nearrow \; KO \\ CO \nearrow \\ HOSO_3, HO \end{matrix}$$

Ex It Burns with a clear, pure blue flame in air - quietly i.e. without explosion.

CO - by mixing with 1/2 its volume of O, combines to form CO_2 - without increasing its original volume

Analysis of CO 4 volume.

CO_2 is composed of - 2 vols of O, combined with 1 vol. C. to 2 vols of CO_2. CO is therefore composed of

1 aeq = 1 vol O = 0.8292
1 " = 1 " C = 1.1056
one aeq = 2 " CO = $\frac{1.9348}{2}$ = 0.9674

It is Colorless + transparent - + highly poisonous. by being absorbed by the blood - where it does not obey the laws of diffusion which other gases there, + can not be removed from it without the greatest opposition - the many cases of Charcoal poisoning - are the result of the action of this gas.

Poisonous Properties

By burning in O - it is converted into CO_2 - which whitens CaOHO - the ordinary test for CO_2

Exp.

A mixture of CO + O will explode but very weakly – can be held in the hand with-
Exp out danger. The reason is that the combustion does not take place rapidly but travels slowly from particle to particle – (With H+O – however, the combustion plants itself fort with a velocity of 34 Meters ~~of~~ in a second).

It is neither an acid nor a base. forms few compounds. With Cu_2Cl, CO will ~~with~~ be taken up – $Cu_2\begin{matrix}Cl\\CO\end{matrix}$
Exp This is an important property. for we make use of it
absorption to absorb the Gas. + of separ-
by Cu_2Cl. ating it from other gases.
It can unite directly with Cl – only however in sun light –

The resulting compound is a Carbonic acid in which one atom of O is replaced by Cl - viz $C\{^{O}_{Cl}$.

With Sulphur it can also unite - uninteresting -

Brought into contact with Hydrated KO - at high temps. converted it has the property of being into form. converted into formic acid ic acid. (forming formate of Potassa)

This CO - brings us directly upon the ground of Organic Chemistry -

Formic Acid.

By boiling this acid with strong oils, it distills over - + forms a liquid - colorless + transparent, forms many beautifully crystallizing salts. The description &c. belong however to Organ. Chemi.

66.

The CO - is interesting as proving many cases of poisoning - in illy ventillated rooms, where coal fires are burning - with insufficient draft, the - headache - & dizziness, so often felt, are the results of the poisonous effect of CO.

Note

c
b
a

The mode of its production in our fires is as follows.

The air coming in from below into the grate - meets a stratum of glowing coal & is burned completely to CO_2 - passing - from b to c - through another glowing mass of C - it is reduced to CO - which if sufficient Ox - be present above c - is passed off - unconsumed - or thrown out, in greater or less quantity in the room.

As dangerous as this gas is in our houses when incautiously generated — just so advantageous does it prove when consumed. In large Iron forges &c. where the furnaces (high blast furnaces) are supplied with alternating layers of coal + Ore. the amt of CO formed in passing through this great thickness of Carbon is immense, + it is one of the most useful of modern contrivances, to economize this unconsumed gas. by leading it through the furnace in such a way that it can be again consumed. The gain in *heat* — is as much as that from the coal itself —

The use of this CO as a combustible.

671

Lecture 73rd

Composition by volume of Light & heavy — Carburetted Hydrogen.

C_2H_4

1 vol C = 0.8292

4 vol H = 0.2768

2 vols C_2H_4 = 1.1060 / 2 = 1 vol C_2H_4 = 0.5530

C_4H_4

2 vols H = 0.1384

1 vol C = 0.8292

1 vol C_4H_4 = 0.9676

which equals the sectional sp. gr.

It can be shown from the "Combining heat" of CO — that more heat is lost by the loss of CO than is economised from the combust. of the Coal

In large manufactories of course this gas is lead from the furnace by perforating air tubes — & is then again consumed by blasts

Oxalic acid. $C_2O_3.HO$,

Formed by allowing finely divided Na (better a mixture of Na + sand) to act upon a stream of CO_2 - **Manufacture**

It exists in plants + animals. In the excretions of animals somewhat small quantities — (Occurrence in Nature)

In manufac. ~~for~~ it is produced from the plants containing it. Obtained as $KO C_2O_3$ & then obtained pure by the following process -

$$KO\,C_2O_3 + CaOHO = CaO\,C_2O_3 + KOHO$$

$$CaO\,C_2O_3 + SO_3HO = CaO\,SO_3 + C_2O_3HO$$

Better - obtained - by boiling sugar with NO_5 many side products are formed by the process but all are easily decomposed - except Oxalic Acid which is stable.

673

This process too - is itself set aside by yet a better one viz: by letting a strong alkaline body (KO HO or NaO HO) act upon wood fibre & by continued boiling till the whole has reached a syrupy consistancy, then the Oxalates are drawn out by HO - & the salts are then again treated as above to obtain the crystallized acid.

Manufacture

It crystallizes beautifully in large white ones - a strong acid taste - it takes up 2 atoms of HO in crystallizing, upon boiling (100°) it gives up one atom & the equivalent formulae C_2O_3 HO.

By heating to high temperatures it is completely volatile — burning to $HO + CO_2$ — Volatility Etc

By cautiously heating this acid — it is partially converted into Formic acid. Note

The salts are stable + many crystallize beautifully

If we heat an oxalate whose base can form a Carbonate — we obtain a Carbonate :— See

$CaO\,C_2O_3 = CaO\,CO_2 + CO$ Behavior of the Oxalate

Note

If the base cannot form a Carbonate we obtain

$Al_2O_3, C_2O_3 = Al_2O_3 + CO + CO_2$

Again. It can under conditions act reducing — with metallic oxalates — Reducing Power see — $AgO\,C_2O_3 = Ag + 2CO_2$

These three kinds of beha-

675

vious are of the greatest importance in the labratory & characterise decidedly C_2O_3

Ex

If we heat Oxalic acid with a stronger acid it is completely decomposed into $CO + CO_2$ & is not blackened. an important behavior in testing.

Treatment with a strong acid

Test

Another test is the heating of $AgO\ C_2O_3$ - when it burns with an explosion - leaving behind metallic Ag.

It is a good reducing agent - it will take from many metallic oxides Au &c &c. their Oxygen & from CO_2 two atoms leaving the metal in reguline form behind. as with Ag.

Besides these oxides of Carbon - there are many others of which the treatment belongs to Organic Chemistry.

C + H.

The number of these compounds is so immense, that we can only treat of the most important ex: (those that occur in nature) - They form the Hydro Carbons especial study - (themselves & their derivatives) of Organic Chemist. There is a certain arithmetical regularity in these compds. & Many rows or isomeric kinds are known - the following are illustrations

$C_2 H_2$ = Metheline
$C_4 H_4$ = Etheline
$C_6 H_6$ = Propyline
$C_8 H_8$ = Butylene
&c &c

} this isomery does not only go this far but far above the 60's.

These bodies have exactly the same relative constitution, & differ from one another only in being, more dense than than one another, as the row is followed downward. Hence they become specifically heavier &c. & proceed from gaseous to liquid & from liquid to solid bodies as we follow the row downward. Another isomeric row: is the following.

$C_2 H_3$ = Methyl	another equally
$C_4 H_5$ = Ethyl	important
$C_6 H_7$ = Propyl	row which
$C_8 H_9$ = Butyl	posses the
$C_9 H_{10}$ = Amyl.	property of acting just

as <u>organic metals</u> - forming oxides, salts etc.

again: C_2H_4 = Hydryde of Methyl
C_4H_6 = " " Ethyl
&c &c

Of these compds. we can form C_4H_4 & the compd. C_2H_4 directly & they both occur largely in inorganic nature - & have much influence geologically - Often by decomposition of Animal & vegetable substances under H_2O - these compds (& particularly C_2H_4) is formed - If the bottom of stagnant ponds &c. is stirred with a stick, bubbels of gas arise in abundance which consists, in mass, of C_2H_4 - & can be collected as the figure represent - by means of a funnel

C_2H_4 in Swamp

679

The dreadful accidents which occur in Coal mines etc - are due, to the accumulation of this gas in the fissures of the mines + its explosion by the light of the miner. In this confined room, it, by the expansion caused by its combining heat - give rise to great tearing + splitting of the rocks.

In Coal mines their Dang.

Where $CaCO_2$ stones are bituminous, + if NaCl is present this gas is formed in quantity - It is not a product of volcanic action - for it cannot exist at high temperatures, being decomposed into CO_2 + HO -

Mixed with air or Oxygen

Ex →

it explodes with excessive violence.

Lecture 74th

We can manufacture this C_4H_4, from Actetic acid or an acetate thus

$$\left.\begin{array}{l} KO, C_4H_3O_3 \\ KO\ HO \end{array}\right\} \begin{array}{l} 2\,KOCO_2 \\ C_2H_4 \end{array}$$

Ex

It is a colorless + transparent gas - of a ~~peculiar disagreeable smell~~ odorless. Which as it has been frequently observed in the regions of decomposing + foetid matters - has been (but incorrectly) supposed to be the cause of Epidemic diseases.

Epidemics from C_2H_4

Has not yet been condensed. Sp. gr. - 0,55314. It burns in air - with a slightly illuminating flame. We most of the Hydro Carbons of this group the gas does not precipitate CaOHO, but the product of

combustion will precipitate it.
Ex showing that the result of
its combustion is CO_2 (+HO)
With Oxygen (1/3 of its vol)
Exp it explodes with great violence
Analyzed with ease; like
all the Hydro Carbons - by
the Eudiometer.

It is a perfectly neutral body
combining neither with acid
nor base - On the contrary, we
can by substitution, entirely
change the nature of the body -
By allowing Cl - to act upon
this gas - in the light (no
Substi- result in the dark) - we can
tuting substitute atom for atom all
Process the H by Cl. thus

C_2H_4 } C_2H_3Cl | C_2H_3Cl } $C_2H_2Cl_2$
ClCl } HCl | Cl, Cl } 2 HCl

Mix the gas (above H_2O) with Cl: + allow one glass to stand in presence of the light, & the second tube in the dark | Ex

The great danger + the fearful accidents resulting in the coal mines, of England — was the cause of an investigation by Sir H. Davy, to find means to prevent the same; + the result — was the invention of the lamp, now, known under the name of the Safety Lamp. Historical sketch | Sir Humphry Davy's Safety Lamp

In an inverted jar — Evaporate some sulphuric Ether | Ex
place the lighted lamp in the ~~burning~~ explosive mixture + show the effect afterward explode or ignite the mixture to show its inflamability

683.

Ethylene = C_4H_4.

C_4H_4 Manufactured in the distillation of Bituminous Coal as an undesired product along with C_2H_4 &c &c,

Pure it can be obtained from $C_4H_6O_2$ — by mixing it with a great excess of $HOSO_3$ — ($1.C_4H_6O_2$ — + 4 to 6. $HOSO_3$) The mixture is poured into sand — to prevent excessive effervescence.

$$\left.\begin{array}{l} C_4H_6O_2 \\ HOSO_3 \\ HOSO_3 \end{array}\right\} \begin{array}{l} C_4H_4 \\ HOSO_3 + 2HO \\ HOSO_3 \end{array}$$

Analysis

Comp. by Volume. It requires for complete combustion to CO_2 + HO, 3 vols. O. + gives 2 vols CO_2 + HO — Hence it is composed of

2 vols H = 0.1384
1 vol C = 0.8292

1 vol C_4H_4 = 0.9676 Sp. gr.

from which the atomic form. can be est'd.

It is neither an acid nor a base – but unlike C_2H_4 it Can combine directly – For example – with Cl. Bring Cl gas into a Eudiometer of C_2H_4 – a compd – + Ex a liquid compound is formed. It takes place in light or in the dark – a pleasant, ethereal smelling fluid, of a wine yellow ft color. If burned ~~with~~ in the air – it burns with an exceedingly bright flame – which renders it an a dmirable substance for illuminating gas –. Mixed with ~~xxx~~ 3 ~~its~~ volumes of Oxygen it Ex explodes Upon, inflaming, with greater violence than C_2H_4 – Shattering even strong glass vessels.

~~6~~85

By mixing with Cl in excess ~~it~~ & inflaming - Carbon is separated & HCl is formed - $\left.\begin{matrix} C_4H_4 \\ 4Cl \end{matrix}\right\}$ 4C + 4HCl.

Ex →

It forms the foundation of an immense number of compounds, known in organic chemistry.

With Conc. SO_3 in Excess - upon constant shaking - there results

Manufacture of $C_4H_6O_2$

$\left.\begin{matrix} C_4H_4 \\ 2HOSO_3 \end{matrix}\right\}$ $C_4H_6O_2$, $\begin{matrix} SO_2 \\ SO_3 \end{matrix}$

If upon this compound we allow a strong base KO, &c it unites with the acid & sets the compd $C_4H_6O_2$ free - this compound when examined proves to be none other than the ordinary subst. known as Alcohol

Alckohol = $C_4H_6O_2$ Is the intoxicating principle of all our so called spirituous drinks. It has all kinds of use in the arts – in manuf.. in science & in medicine –
It results as a product of what is called Alcoholic fermentation – viz –: Grape sugar when brought into contact with a ferment – (an organic substance in decomposition, which carries over or transmits this decom. process to the body with which it is brought into contact) is decomposed into CO_2 & into alcohol & thus it is generally obtained. See: –

$$C_{12}H_{12}O_{12} + \text{ferment} \left\} \begin{matrix} C_8H_{12}O_4 = \left\{ \begin{matrix} C_4H_6O_2 \\ C_4H_6O_2 \end{matrix} \right. \\ 4CO_2 \end{matrix} \right.$$

Lecture 73rd

Alcohol = $C_4H_6O_2$ $\left\{ \begin{matrix} C_4H_5 \\ H \end{matrix} \right\} O_2$

The various sugars possess the property of passing into one another by taking or giving up HO atoms. The change of the sugars into $C_4H_6O_2$ by fermentation depends upon the presence + growth of microscopic plants. If we shut off all the air from the surface the sugar solution will never ferment per se. The presence of a 'ferment' is generally the case as it facilitates the process of decomposition, in that, the vegetable organisms generate themselves in the substance to be fermented.

Fermentation

It is a colorless thin fluid of a specif. grav. when absolute at 15° = 0.7996 - Burns with a bluish & little lighting flame. Ee

It is universally used as a solvent for great classes of bodies - such as the Alkalies, Iodine - Ethereal Oils, Resins - & many uses are made of its solvent power as a mode of separating Salts &c, in Inorganic Analyses.

By distilling $C_4H_6O_2$ with Conc. SO_3HO, there results C_4H_5O, Ether as it is called. Ether $\left.\begin{matrix}C_4H_5\\C_4H_5\end{matrix}\right\}O_2$

This substance - is likewise a colorless fluid - & volatilized shown at 53° It is a solvent of many substances for which $C_4H_6O_2$ is not Ee

689.

$C_4H_6O_2$ can by allowing Oxygen to act upon it under certain conditions Acetic acid + HO is formed.

Acetic Fermentation

$$C_4H_6O_2 \left.\begin{matrix} \\ \end{matrix}\right\} C_4H_4O_4$$
$$O_4 \quad 2HO$$

This is called acetic fermentation, & demands the constant presence of Oxygen & a somewhat higher temperature than $C_4H_6O_2$ ditto.

It is formed by fermentation of $C_4H_6O_2$ in presence of Nitrogenous animal or vegetable decomposing matter – the process is the following:–

$C_4H_6O_2 + 2O = C_4H_4O_2$ = aldehyd.

$C_4H_4O_2 + 2O = C_4H_4O_4$ Acetic acid.

Ordinarily known under the name of Vinegar.

$A + B + C$

$nA + n_1 B + n_2 C$

$\alpha nA + \alpha n_1 B + \alpha n_2 C = M$

$\alpha n C + \alpha n_1 H + \alpha n_2 O = m$

$\frac{\alpha n C}{C} + \frac{\alpha n_1 H}{H} + \frac{\alpha n_2 O}{O}$ Formula

$\frac{\alpha n}{\alpha n} + \frac{\alpha n_1}{\alpha n} + \frac{\alpha n_2}{\alpha n}$

Is the general formula from which we derive the equivalent formulae of Complicated Compd's.

Sugar. Acetic acid. Same comp.

$40 C\%$ $\frac{40}{6} = 6.6 = \alpha n$

$6.7 H$ $\frac{6.7}{1} = 6.7 = \alpha n_1$

$53.3 O$. $\frac{53.3}{8} = 6.66 = \alpha n_2$

$\frac{6.6}{6.6} + \frac{6.7}{6.6} + \frac{6.6}{6.6}$

$1. + 1 + 1$

$C. + H + O$ Empiric formulae for Acetic acid & Sugar

If we try with acetic acid to substitute the metal for H. we find that we can substitute 1/4 of an atom of H. with 1/4 th of a ditto of K. as we know of no fractions of atoms in Chemistry we must multiply the formulae for acetic acid by 4 + we then have $C_4 H_4 O_4$. With grape sugar we can substitute 1/2 of an atom of K or some other substance for a ditto of H - hence we use the form. $C_{12} H_{12} O_{12}$ Thus the formulae of all the immense numbers of Organic Compounds have been derived

Obtaining Formulae

69[3]

Of another kind is the so called Molecular formul. | Molecular – Formulae

From the fact that gases expand equally for changes in temperature &c. the conclusion is drawn that in equal room (volume) of gaseous bodies – an equal number of atoms or groups of atoms are contained –

O	H / H	HO 9	Cl ~~35.5~~	H Cl 36.5

Facts, however, show that the greatest volume formed

8 8 / 8 8	1 1 / 1 1	9 / 9	35.5 / 35.5

by the combination of any substances, is that taken in by H Cl = 4 volumes – For –:– 1 vol O = 8 parts by weight – combines with 2 vols of H (or 1 pt. by weight) to form 2 vols of HO

2 Cl + 2 H ~~ar~~ vols, combine to form 4 vols HCl –

And this is the greatest volume that is formed by any com-

694

bination –

The ~~atom~~ Molecular formula – would then, be, the formula which expresses the relative weights of equal volumes of various compds. & that volume would be 4 vols – which is the greatest volume formed.

So ~~with~~ HO – which forms from 2 vols H + 1 vol O – 2 vols HO – the formula must be doubled to comprise 4 vols. thus giving us an expression for the presence of an equal number of atoms or atom groups with HCl &c

9

HO 2 vols

9	9

HO 4 vols

Would make the molecular formula of HO = 18 instead of 9 –

Lecture 76th.

These molecular weight of the various Elements &c— are proportional to their sp. gravities. If we know then the sp. gravity - of a substance we can reckon its Molecular weight. thus:—

Sp grav O : Sp grav of M : Molec. wght O : M
1.105 &c : S = 32 : M or $\left(\frac{32}{1.105}\right)S = M$.
$= (28.94)S$

for all cases – for Example.
Specif Gravity of SO_3 vapor— is 2.764. Its molecular form is.
2.764 × 28.94 = 80. = Molecular wght of SO_3 which is in the relation of 2 to 1 – with the Equivalent Weight. making the formula S_2O_6. In Organic Chemistry this method cannot be too highly prized – whereas – it is valueless in Inorganic Do.

Value &c.

696

Sp. grav of $C_4 H_4$ = 0.94

mult by = 29

Ethelyne 28.13

C_4 = 24 $C_4 H_4$ = 28.

H_4 = 4

$C_4 H_4$ = 28

$C_4 H_5$ Sp-grav = 2.002

Ethyl - multiply = 29.

$C_4 H_5$ 58. molec. weight

24+5=29 eq. weight ↑

So that the molecular weight is twice that of the Equiv-do- & the formula would be written $\left\{\begin{matrix} C_4 H_5 - \\ C_4 H_5 - \end{matrix}\right\}$

Worth of the new Theories

The "new theories" as they are called - in the comparatively simple domain of Inorganic Chemistry - only produce more complication without giving us a clearer idea of the positioning of the atoms

It was formerly generally conceeded, to call the lowest or smallest weight (relative) of an Element, entering into a combination the atomic wght; but the ground is too unscientific to be definitely held.

The lowest wght of a subs. in Compd. the atomic Wght

It is now generally agreed (since Berzelius' time) to call ~~call~~ these numbers atomic weights which correspond to the law of the relation of Specif. Heat + Atomic Weight

Atomic Wghts.

Some products vary from those given by the majority of the Elements but vary in simple multiples (the normal number is 3.+) The equivalents (thermic) taken as the atomic weights of all these would then be ~~[illegible]~~ ~~[illegible]~~ halved.

698

The use of C_4H_4 &c is as illuminating material, Produced by bringing the gas to a high temperature - where chemical & physical changs show them.

Flame principle of.

A body being heated - is from physical principles constantly taking up & giving out light. It is by no means all the light rays which are given out - that we can perceive - for rays beyond the red - & beyond the violet are imperceptible. As the temperature is increased first the red (heat) rays, are perceptible to the eye & then as the temperature rises - the various colors to violet become perceptible - the mixture of all with solid bodies making <u>white light</u>.

but there is a range of rays that extend far beyond the violet & far beyond the red invisible to us.

The amount of light which one & the same body gives out depends 1st upon the temperature & 2nd upon the nature of the body. viz: the Emission of light being directly proportional with the capacity the body has for absorption. By lowering the temperature the power of a body to absorb light is lessened. by elevating it its power of absorption is increased & consequently also its power to emit light. the two standing in direct proportion.

We use universally the Hydro Carbons to produce light. & especially the C_4H_4 group.

700

The products of Combustion however – are CO_2 + H_2O – both of them transparent gases of no, or very little Emissive power. The cause of the intense illuminating power of this Group is that they are decomposed at higher temperatures into lower Hydro-Carbons + Solid Carbon – a perfectly intransparent – black body of immense absorptive power in higher temperatures – + a correspondingly great Emitter of light. It is separated in an impalpably fine state + passing through the intensely heated non luminous flame of gas – is heated to White heat + emits its powerful light.

Cause of illuminating

Same can be brought about by C + CaCO

Ex

Lecture 77th

To prove that the emission of light is dependent upon the temperature have two flames of high + low temps — (i.e. Bunsen's Lamp + $C_4H_6O_2$) + place in each, pieces of Pt wire — one will be brilliant white the other very weakly red. That it is dependent upon the nature of the body place in the same flame a bead of 'Borax' — + a 'Pt wire'

Exp

Prove that the Emission of light is depend upon Temp

Exp

So these Hydro-Carbons in the heat of a combining gas are decomposed in various ways — depending upon the presence or absence of sufficient Oxygen — thus —

$C_4H_4\begin{cases}C_4 \rightarrow \\ 4HO\end{cases}$ or $C_4H_4\begin{cases}C_2H_4 \\ 2C\end{cases}\begin{cases}C_2O_2 \\ 4HO\end{cases}$

or $C_4H_4\begin{cases}4CO_2 \\ 4HO\end{cases}$ = when Oxygen in abundance is present.

Exp

c 702 b

a

Shape of the flame

The gases per se stream forth + became regularly wider + wider – but when inflamed, ~~they~~ the flame assumes a conical form – for the products of combustion comming into contact with cold air – as they recede from the opening – ~~make~~ become less + less luminous + finally end in a point – producing on the whole – the conical form.

Candle light Decomp.

With a candle, the wick acts as a cappillary tube (or rather as a series of them) bringing ~~to~~ into contact with the ~~heated~~ flame the liquid tallow – + volatilizing it – + these – products are Hydro Carbons, + chiefly $C_4 H_4$ – the process of the combustion of which we have just explained.

In a flame we can distinguish very different portions — & these divisions are of the greatest importance to the Chemist. 1st A dark hollow space in which the cold gases streaming out from the lamp – are heated – & H_2O might be vaporized. Then 2nd the flame mantle – composed of the inflamed & combining gases made luminous by the numberless particles of solid C. in it. And 3rd the film at the surface & top of the flame – composed of the more or less luminous products of combustion – heated from their production in the mantle. 4th – a small portion at the lowest part – where the outstreaming gases meet a supply of cold air.

Divisions of an Ordinary flame

704

Exp. By holding a cold + bright surface of metal - obliquely + squarely across the flame - a vertical - + a cross section of the flame can be obtained ¹ ² - the two appearing - as indicated by figs 1 + 2.

Show the sections + temperatures of various parts by wires

By placing a wire in the top - of the flame, it is made to glow - as we lower it. it glows without + within - but in the centre is an unheated part which is dark - for it is in the cold part of the flame. These various parts of the flame are used - for the various purposes of Reduction - + oxidation - + the aparatus serving its the purpose of performing the operation is termed the blowpipe -

The Blowpipe

Bunsen's lamp – has now entirely supplied the place of the Blowpipe – for convenience. Bunsen's Lamp

The gases are lead from 3 or more openings into the bottom of a tube + air is also admitted from openings – at the bottom – + the gas is consumed without – C being separated, + the flame is non-luminous.

We can distinguish the fallowing parts

1st A lower Reducing flame on the ~~outer~~ inner surface of the flame mantle (a)

2nd An upper Reducing flame at b. 3rd A lower oxidizing flame at c – 4th A higher oxidizing flame at d – + 5th A fusing flame at e – the hottest part – in the centre of the flame mantle.

Ex

The various metallic oxides give various colors to the glasses - so that by bringing a bead of borax glass - in these various parts of the flame Reducing or oxidizing, we can produce the various tests in the most perfect & convenient way by the Bunsen Lamp.

How to Analyze the gases from an Oil-flame

By bringing a tube of glass connecting with a collector a and an aspirator b into connection with an oil flame (into the dark portion - the products obtained from the volatilization of the oil can be collected in the collector a - & analyzed.

C + S -

The only c'd of C & S is CS_2 -
Formed by leading the vapors of S over glowing Coals - & the substance is formed by direct combination.

A flask vessel is filled with pieces of coal - & heated from below & from above - pieces of Sulphur are dropped in - the CS_2 distills over & is collected over HO - the box a is to collect the Sulphur.

When Chemically pure it has no odor - but impure its smell is disgusting - when concentrated. it is however not unpleasant.
A colorless liquid - Sp. grav 1.271 at 15°
Does not unite or mix with HO but sinks to the bottom without mixture.

Burns in air with a pure blue flame - to CO$_2$ + SO$_2$ Decomposed by leading through a hot tube of porcellain it is decomposed into CO + SO$_2$ - volatilizes easily at ordinary temperatures - & is converted into colorless gas.

Properties

It is a very important body to the chemist as a solvent - for many bodies which HO or C$_4$H$_5$-OHO does not dissolve - viz: Sulphur. Phosphorous - & also of Iodine - & Bromine. In analytical operations it is invaluable as a separator of Sulphur - of solution - particularly also of Iodine & Bromine.

C + Cl re.

Cannot be formed directly but must be produced by substituting Cl for H in some of these Organic Compds called Hydro Carbons. - forming the most interesting of all the processes of Organic chemistry - of theoretical value - but practically of none.

Lecture 78th

By this means - (viz. substitution) we are enabled to manufacture a whole row of C + Cl compds. Or - by setting out a vessel of CS_2 to a high temperature & passing over it a stream of Cl - A simple decomposition into CCl_2 + S_2Cl are formed & distil over - S_2Cl is decomp. by leading through alkaline solutions

Cl

CS_2

CCl_2 is a colorless fluid - does not mix with HO - & has a pleasant - peculiar aromatic smell - Boils at 78 °C.

Ex Is difficultly inflammable With HO + alkalies it is not decomposed. From its specific gravity its molecular weight can be determined.

C_2Cl. C_1Cl, C_2Cl_3 $\underline{CCl_2}$ -
These are the compds of C & Cl which are known but none are interesting here - but in Organic Chemistry - $\underline{C_2Cl}$ - is a solid - looking & smelling like Camphor - Not decomposible by HCl or Alkalies Soluble in $C_4H_6O_2$ - Boils at 200°
$\underline{CCl}$ - a colorless liquid - boiling at 120° - & equally indecomposible by alkalies

$\underline{C_2Cl}$

C + Nitrogen

C_2N = a compound radical & called Cyanogen

$\underline{C_2N}$

It can be formed directly - but a third body with which it can combine - must be present - viz: mix C & KO - in the slightest heat & lead N - over it - KC_2N + CO are formed (formed often in Blast furnaces)

Again, By treating Oxalate
Manufacture — of Ammonia with a great excess of SO_3 (50 or 60 times)

$NH_4OC_2O_3$ } C_2N ↗
+ - SO_3HO } $4HOSO_3$ aq.

C_2N is given off + HO formed. By Heating HgCy it ~~is~~ falls apart into Hg + Cy ↗.

Properties — It is a colorless - transparent gas - (also a black solid - allotropic modification) which possesses a smell like bitter almonds - + appears to be very poisonous. Can be condensed at -34°F. It burns with a beautiful violet red flame to CO_2 + N - The spectrum is not
Spectrum — that of C, but a peculiar one of C_2N - as it does not decompose.

Mixed with O – it explodes upon inflaming – with a sharp noise. Eq.
It is soluble very largely in H_2O + $C_4H_6O_2$ –
It acts to decompose H_2O after some time – forming – C_2H_6 + O ??
Composition by volume of this gas is to be found in chemical works; & some considerable deviation is found from the ordinary rule. (i.e. 2 vols – C vap – to 1 of N – to 1 vol C_2N ?) – It is called a compound radical like NH_4
It acts like an element & indeed like an element of the group of Cl – Br I &c
I could properly regarding its compds – have treated of it when we treated of those bodies.

714

Combustions can be carried on (just as with Cl &c) in the gas (i.e.) generate it by a tube of $HgCy$)-

Ex

HgCy Na

$H + Cy$.

Formed like HCl - by treating a Comp'd of Cy - with a Hydrated Acid. It is the surest - & quickest of all the poisons known to chemistry - Is manufac. again thus -

Manufacture

$$\left.\begin{matrix} KCy \\ KCy \\ 2HOSO_3 \\ Aq \end{matrix}\right\} FeCy \quad \left\{\begin{matrix} 2HCy \\ 2KOSO_3 \\ FeCy \end{matrix}\right.$$

The Aq - representing the process to take place in presence of an excess of HO.

It is made concentrated by desiccating with CaCl + distilling over in a vessel surrounded by Ice. If not handled with excessive caution it may explode - when condensed per se - causing fearful accidents from its deadly poisonous properties. In this latter relation it is to be remarked that it is the most fearfully, + deadly poisonous body known.

Dangs of Explosion.

Poison.

It burns with a slightly colored flame - to $CO + CO_2$

A dove allowed to inhale the fumes - for a few second a is killed - even when the same, are given off from the concentrated acid at 0°C. when, it can be but trifling in quant.

Dove Exp

There is no chemical antidote for this substance - the treat-

Antidotes — ments - viz:- application of cold &c to the spine - are founded entirely upon medical grounds & are utterly useless.

By bringing HCy into contact with Metallic Oxides, it acts

Cyanides analogous to Chlorides &c &c — precisely like HCl - we obtain a Cyanide + HO -

$KO + HCy = KCy + HO$.

So that with HCy - we can produce all the Cyanides -

The analogy goes so far that - even with reagents - we obtain similar precipitates - With $AgONO_5$ - - we obtain - a white, curdy precip. of $AgCy$ (only soluble in conc.

Tests & Reactions — SO_3) but also in NH_4O like $AgCl$ - So also a precipitate with $PbOA$ -

Can be however distinguished by its smell. (bitter almonds) then a peculiar reaction - which will soon be mentioned - by which a peculiar deep blue precipitate is brought about - of what is called - Berlin or Prussian - blue - owing to the property of the substance C_2N to form a remarkable lists of Double Salts -

---

Lecture 79th. If we add to one portion of a solution of H Cy KO + to another a solution of ditto a mixture of an FeO or Fe_2O_3 salt - & then after dissolving up the precipitated Iron salts - in an acid, add the two mixtures - we get a beautiful dark blue precipitate of Ferro-cyanide of Iron, an especial test for Iron. - The salts of H Cy with K. with Na, Ba &c are of the nature of the Oxygen compounds of these bodies - (viz: KO NaO &c i.e. they are bases) again: Fe Cy, $Co_2 Cy_3$ &c &c - play the parts of acids - & these two classes of compds, unite with one another to form double salts.

$\left.\begin{matrix} KCy \\ KCy \end{matrix}\right\} FeCy + aq.$ Is a much used substance, especially in manufacturing, (<u>coloring</u>)

Manufactured largley - from Nitrogenous animal matter - (Horn - Leather refuse - old cloth &c) - by smelting them with KO - in Iron retorts, with addition of Iron filings - & then dissolving out the salt with HO. From this double salt we can form a whole row of other salts - many are beautiful in Color - & with Metallic Oxides - form all insoluble ditto's - Hence this salt is much used in the

Import. in Technic.

Manufacture.

720

Ditto → $3\left(\begin{matrix}KCy\\KCy\end{matrix}\right\}FeCy\left.\right)$ $\left.\begin{matrix}\\ 2Fe_2O_3\,3\bar{A}\end{matrix}\right\}$ $6KO,\bar{A}$ / $2Fe_2Cy_3\,3FeCy$

A beautiful Prussian blue. the Ordinary test for Fe.

Test for Iron

By leading Cl into the Ferro-Cyanide of Potassium — another Compd — Ferri-Cyanide of K is formed — thus

$2\left(\begin{matrix}KCy\\KCy\end{matrix}\right\}FeCy\left.\right)$ $\left.\begin{matrix}KCy\\KCy\\KCy\end{matrix}\right\}Fe_2Cy_3$ Cl

Ferri-Cyanide + prot. generally

Called Red Prussiate of Potash. Important as a principle in Coloring &c.

$$2\left(\begin{matrix}KCy\\KCy\end{matrix}\right\}FeCy\Big) = \left.\begin{matrix}KCy\\KCy\\KCy\end{matrix}\right\}Fe_2Cy_3$$

Cl → Cl

With Sesqui Salts of Fe, it gives a light, instead of a dark blue, precipitate.

C_2N + Cl, & Brol.

By treating HgCy with Iodine there results:—

Hg Cy } Hg I.
I I } Cy I.

So also we can produce the Compd Cy Cl, a gaseous body very Poisonous.

Ditto with Br. With Cl there is formed another (or rather ~~polymeric~~ isomeric variety) Compd. viz:- a union of Cl- with that which we called- the allotropic modification of C_2N. From this we can form the Oxyd of C_2N- by simple decomposition

~~Cy3~~ Cl_3 } $Cy_3 O_3$, 3HO ⟵ $Cy_3 O_3$ 3 HO
HO
HO } 3 HCl.
HO
eq-

Forms a whole list of salts
The products of its decomposition are remarkable. by distillation, it forms an isomeric compd. of entirely different properties – for it is Monobasic.

$$Cy_3O_3 . 3HO = \begin{Bmatrix} HO\,CyO \\ HO\,CyO \\ HO\,CyO \end{Bmatrix}$$

This latter decomposes itself in few minutes – into an utterly indifferent body – neither an acid nor base – inodorless &c but still isomeric – So that we have three isomeric modifications of Cy + O.

By glowing KCy in air – it takes up O. + is converted into Cyanate of KO. partially – KCy remaining over can be separated as it is insoluble in $C_4H_6O_2$. KOCyO is soluble.

Therefore by treating the substances with $C_4H_6O_2$ we can separate them.

CyO in HO solution forms instantly $NH_4O\,CO_2$

$2KO\,CyO$ } $2KCl$

$4HCl$ } $2NH_4Cl$ ↗

aq } $2CO_2$

By the evaporation of a solution of $NH_4O,(C_2N)O$, this peculiarly Animal excretion – the principle of the Urine – is obtained – viz:– NH_4O,C_2N,O. =

Urea its formation from $NH_4O(C_2N)O$

$C_2H_4N_2O_2$ = Urea –

the Empirical formula for both being alike – long white crystals – bitter taste – soluble in HO + in $C_4H_6O_2$ – Solution in HO perfectly neutral.

724.

C_2N + Sulphur –

Formed precisely like CyO. by smelting KCy with S – in the air – viz:–

$KCy + 2S = KS, CyS.$

Sulpho Cyanogen & Salts –

& drawing out the Sulpho Salt by $C_4H_6O_2$ as before. It forms a whole row of double salts like the corresponding Oxy-Salt. viz:–

~~$HgO\bar{A}$~~ } $KO.\bar{A}$
~~$KS.CyS$~~ { $HgS.CyS$

So also with all the metallic Oxides – gives beautifully colored salts –

With Salts of Fe_2O_3 – it gives a beautiful reddish brown

Test

coloration – a fine, delicate test for either Fe or CyS.

The salt HgS CyS. forms Pharaoh's as shown before by the Serpent. following reaction; i.e.

HgO Ā } KO Ā ⟶
KS, CyS } HgS. CyS.

Gives a most peculiar appearance upon inflaming it – Its products of combustion – take up immensely in bulk – the combustion being accomp. with a bluish flame. This swelling in bulk. has lead to the use of this Substance to form the toy – Called – Pharoah's Serpent. well known to all,

Lecture 80th

COS Oxycloride of Carbon. Found in Nature in many Mineral Waters - Can be regarded as a CO_2 in which 1 atom O is replaced by S, viz: $C\begin{cases}O\\S\end{cases}$. Decomposable by Water - Burns with a pale blue flame. Can be lighted by a glowing stick like O. By Electrical decomp- it separates Sulphur but does not alter its volume.

Boron - Bo.

Comes much more abundantly in nature than was previously supposed namely in Sea Waters as BO_3 & in many fucoids - as BO_3 - As a constituent ~~of~~ of Rocks.

it is not found. But it forms ~~many~~ a constituent of many Minerals. e.g.: | Die Minerals
Boracit $3MgO.4BoO_3$ -
Stassfurtit $2MgO, 2BO_3 + xaq$.
Larderellit. Borocalcit. Tinkal
Sassolin, (Tourmalin?).
Its generation (principally as Sassolin) - is connected with volcanic appearances.
Boron - is obtained from BO_3 by smelting with an excess of Na - It is a smutty green, completely amorphous mass. | allotrop. i.e. modific.
By smelting with Al - it is obtained crystalline - (like Carbon) corresponding to graphite - Quadratic Octahedra)
The latter is difficultly inflammable - the ~~latter~~ former much more easily.

By bringing Bo. into contact with KoHO - it is upon fusing - oxidized :- viz:

Bo
HO KO
HO KO
HO KO } 3(KO BoO3)
3H

By sprinkling some of the finely divided Bo in a vessel of Cl gas it is inflamed.

Exp

It possesses the remarkable property of combining directly with Nitrogen

It shows great analogy in its properties & its compds to Carbon -

From the impure Sassolin – $NaO\ 2\ BoO_3 + 12\ HO$ is manufactured by neutralizing the BoO_3 with $NaO\ HO$. – & allow to crystallize. It can then be separated & obtained crystalline, by treating with a stronger acid (SO_3) & evaporating to sufficient concentration.

It shows a peculiar action toward Test-paper. (Litmus)

If to a hot solution of Borax, colored blue by litmus tincture – Conc SO_3 be slowly added at first no reddening takes place for the SO_3 drives out the BoO_3 & unites with the NaO – & the free BoO_3 has no power to redden it – Even though an acid. It is only when an excess of SO_3 has been added that a reddening sets in.

Ex BoO_3 & litmus paper – Ex litmus –

BO_3.
It crystallizes in large crystals – by glowing can be obtained anhydrous – for it is like PO_5 a fixed nonvolatile acid. & forms a colorless & transparent glass.
In the flame – it gives a very weak green flame but must be highly heated.
When melted it can like PO_5 – be drawn out into threads –
It shows very singular relations with bases.
1 Base to 2 Acid – 3 Base to 4 Acid – 4 Base to 4 acid. &c
The principle salts are the following:–
Borax ($NaO, 2BO_3 + 10HO$.)
a very important salt, universally used in the arts, & domestically.

$AgONO_5$ gives with BoO_3 salts
a white Precipitate — Efe
BaCl - gives likewise a white
Precipitate
Both, however, of these precipitates - are soluble in HCl.
There is however one peculiarity - by which the Borates
are distinguished from all
other salts - viz:—
When we separate the BoO_3
it has the property of acting
upon some vegetable colors
like a base - i.e. With Turmeric
Paper it gives like the alkalies -
a brown color when dry - Ell
again - upon separation - from
combination by a stronger
acid - it volatilizing in slight
Quantity gives a pale green
Color to the colorless gas flame.

Lecture 81st

The spectrum of BoO_3 is characteristic, & easily detectable the lines lie within the range of green. This light can be brought about by ex-

Ex- posing a pearl of BoO_3 to a sufficient heat to produce volatility of the BoO_3 - & its salts treated (in the lowest part of the flame), with conc. SO_3 -. by this means we can detect the smallest traces of BoO_3 & is the best reaction.

Flame Testing for BoO_3

As all the Borates are soluble (nearly) it is difficult to detect this substance quantitatively. That one salt is $MgO\ BoO_3$ - & the salt must be highly heated - which produces the property of insolubility - & as such $Mg_2O\ BoO_3$ - is weighed.

This acid is used to reduce the manufac. of false precious stones - Being non volatile it possesses too - the property of forming a colorless glass which can be colored by metallic oxides. False P. Stones

These salts are used too - to aid the soldering of metals acts as follows: - upon heating ordinary metals Sn. Cu. Zn &c, a coating of Oxide is formed, this coating is dissolved up in Borax - if it be rubbed upon it - & the metallic surface is kept perfectly bright - so that the fusible metal which is to unite the two surfaces Soldering can easily fuse fast to them otherwise the coating of Oxide would prevent the Union

By Reduction with NH_4Cl. NH_4Cl is also used in soldering, but its action is that of reducing - & not as with Borax ~~Oxide~~ a solution of the Oxide formed

On acct of this property of dissolving up metallic oxides - Borax is much used in the Laboratory - to test the different metal. Oxides

Beads of Borax with MeOx. A clear bead of Borax, is dipped into the subst. or solution, & then held in the Oxidizing flame (or R) - on upon solution - the different colors given by the different Metals can be employed to distinguish what metal is present.

B + Cl.

By distilling BoO_3 with concentrated HCl – & we obtain it in traces only – for HO dis~~til~~ ~~til~~ decomposes it. Manufacture

By leading Cl – in a glowing tube over a mixture of BoO_3 + C. – or over BO. It is gaseous.

Do

B + Fl.

By mixing a fluoride with BoO_3, & covering the mixture with SO_3 – we obtain $CaOSO_3$ + $BoFl_3$

$$\begin{array}{ll} & CaFl \\ BoO_3 & CaFl \\ 3\,HOSO_3 & CaFl \end{array} \quad \begin{array}{l} 3CaO\,SO_3 \\ BoFl_3 \end{array}$$

A gas which can only be caught over Hg. Forms aut thick clouds in presence of HO vapor –

736

HCl decomposes it into $BoO_3 + H_3Fl_3$.

$HFl, BoFl_3$ — corresponds to Hydro-fluo-silicic acid.

the manufacture of which will be shown under the head of Silicium.

These salts are precisely analogous to the Oxy-salts where the O is replaced by Fl.

Bo + N.

Manuf. Is a body which is very stable — by leading NH_3 over heated BoO_3 — can be heated to redness without decomposing. $BoO_3 + NH_3 = 3HO + BoN$. A white solid.

Boron demands more careful study than has hitherto been devoted to it.

The Method of Organic Analysis

That is universally followed in Elementary Organic Analyses is the following - A Verbrennungs Rohr - drawn out to a fine end - is filled to a - with the dried + hot CuO - after assurance that the whole tube is carefully dry, glowing with CuO

CuO Cu CuO CuO + subst CuO c b a

then, the tube containing the weighed substance is placed in the end + some allowed to fall in upon the CuO

CuO

(all necessary precautions being taken (vide Bunsen's Method) to prevent action of moisture) some more CuO is then placed upon it + the subst + CuO mixed with this concern [drawing] of Cu, then more CuO is placed upon the mixture - to c, + then at the farther end a ~~cole~~ roll of metallic Cu (first oxidized by heating

+ then given a brilliant metallic surface by reduction in H gas) [coil drawing] like this is placed - if the substance to be analyzed is nitrogenous, if not - it is unnecessary - for it (the Cu) has for its object the reduction of any NO - NO_2 or other Oxides of N which may be formed: at the farther end then the CaCl tube + the KO apparatus - are arranged air tight - ~~the latter~~ both carefully weighed - the heating is then commenced from D towards A - & after an hour or so the operation of heating can be suspended, + the fine end broken off + by an aspirator - dry air lead through in the direction a to d. - the weighings then will give the weight of C + H. for O + N ; other plans are used -

Silicium.

Next to Oxygen – the most widely spread in nature – Has acquired new interest by the late discovery by Wöhler, that like C, it can + does enter into a great number of compds. analogous to + identical with the so-called Organic Compds of Carbon.

The new Interest which Si has acquired.

Next to O, Si is the most wide spread of all the elements. Found every where, except as a constituent of the atmosphere – As a metal however it does not occur upon the Earths surface but as free SiO_2 or as a combination of this acid with the various metallic Oxides – KO, NaO, CaO, MgO FeO + Fe_2O_3 – + chiefly – as $Al_2O_3 \cdot 3SiO_2$.

As Rock Crystal - it comes
quite pure SiO_2 in nature -
sometimes in great large
Crystals of ~~feet~~ one feet in length.
Then it occurs in the va-
rious varieties of Quarz -
zb. Amethyst - Opal various var.
Jasper - Chalcedony - Agate &c &c.
Then as a constituent of
an innumerable number
of rocks & minerals called
Silicates - our most common
& wide spread rocks &c.
Granite - Syenite, Dolerite Basalt &c.

Si Obtained Allotropic

Obtained like Bo - by smelting together a carefully prepared mixture of Silico-fluo-
ride of K with Na & Zinc.
Or by the action of Metallic K
upon Silico Fluorid of K i.e.
$KFl + SiF_2 + 2K = 3KFl + Si$ - amorph.

Possesses great resemblance to metallic Bo - burns in Cl gas with the latter &c. very slightly attacked by NO_5. With KoHO & all the strong alkaline bases attack it with ease -

$KOHO + HO + Si = KOSiO_2 + 2H$ ↑

H_2Si ⟵ Si & H.

A colorless gas which inflames spontaneously in the air & burns - giving off a white vapor. Obtained by the solution in HCl - of Manufacture Aluminium - containing Eq Si in solution.

$4Al + 3Si + 6HCl\ HO =$

$2(Al_2Cl_3) + 3SiH_2 \nearrow + 1HO.$

Lecture 82nd

SiO_2 — Silicic acid.

Comes in foot long crystals as Rock Crystal. & all the silicates contain the acid. & really forms the subject almost of Mineralogy.

The mineral containing the SiO_2 as an acid - is treated - with an acid viz. HCl - & evaporated to dryness. By this

Extraction of SiO_2

means - the acid is separated from its base - & a chloride formed - then - upon again dissolving out in H_2O - the SiO_2 is left behind as an insoluble residue. & the chloride is dissolved up.

$$CaO\ SiO_2 \} CaCl\ 2HO$$
$$HCl\ HO\ \} SiO_2$$

A filter is now only necessary to separate them.

As thus obtained, it is a fine white impalpable powder. There exists two modifications of SiO_2, a difficultly & an easily soluble modification. The first is called the crystalline modification & includes the various varieties of crystalline & semi-crystalline SiO_2. The latter is represented by Opal & its derivatives – as well as by the beds of SiO_2 composed of myriads of the houses of Infusoria.

2 Modifications

Diatomaceae

Only soluble in H_2O in traces, with strong bases – the first modification is insoluble (except at high temps.) – in H Fl, it is likewise soluble forming SiF_2 – Both modifications are insoluble in H_2O, & in acids.

744 The ~~first~~ second modification dissolves even in the cold in KOHO
The ~~second~~ 1st - only dissolves by continued application of Heat.

Properties of the two modific.

The Sp. grav. of first - = 2.3
" " " " second = 2.6
The relics of Diatomaceae belong to the first modification. Rock Crystal when smelted belongs to the second.
To the difficultly soluble modification, Flint belongs -
To the lightly soluble one - Opal belongs.

Rose's Discovery.

Rose discovered that by a smelting of the crystalline modification in the Oxyh. blowpipe - that it is converted into the lightly soluble modific.

Hence many have used the argument - that Granite - could not have been of plutonic Origin - because, by being of a fiery origin i.e. being ejected from the Earth as are modern lavas the Quartz contained therein must have been converted into the soluble modification; whereas it exists as the difficultly sol. one. But - the SiO_2 may have existed dissolved up in the melted rock (or the silicates (viz: feldspar) may be able to take up more SiO_2 into combination when in a melted state), + the point at which the dissolved SiO_2 might crystallize out may be far - far below that at which it melts - hence the argument looses its force.

Geological Adaptation.

745

The only salts of SiO_2 which are soluble – are those of KO, NaO, &c. of the alkalies – If KO SiO_2 is treated with – HCl – & the SiO_2 is separated out in ~~a~~ amorphous state –

Note

Exp

If, however, the experiment is reversed – & the KO SiO_2 is added suddenly – to the HCl – nothing is precipitated & the liquid remains clear, because – the SiO_2 in the moment of its separation is soluble – &

Explanation.

in the second experiment each portion separated finds abundant HCl to dissolve it up – in the first case it does not. by a sufficient time of standing ~~the~~ SiO_2 in the second case separates – by – about an hour or two afterwards.

This property of insolubility in HCl is an important property of SiO_2 - by which we can separate it from all other substances soluble in that medium. We have 3 classes of Silicates viz:-

3 Classes of Silicates

1st Such as are soluble in HCl. i.e. Silicates of the alkalies.

2nd Such as are decomposible by HCl - many Silicate minerals.

3rd Such as are insoluble in HCl & indecomposible by HCl - but which must be fluxed with Carbonated alkali or treated with HFl.

Fluxing

With this 3rd class to analyze the SiO_2 quantitatively we must use the process called fluxing i.e.

~~$CaO\ SiO_2$~~ | $CaO\ CO_2$. Whereby a
~~$NaO\ CO_2$~~ | $NaO\ SiO_2$ Carbonate

of the SiO_2 base is formed + the Silica bound to the NaO.

747

The acid is non-volatile & infusible by the highest temperatures we have been able to bring about.

Besides SiO_2 there is another Compd of Si & O.

With Cl Si forms a corresponding compd of Cl.

Organic Compds of Si

Formed by leading Cl through a glowing porcellain tube in which is a mixture of SiO_2 + C. viz:—

$$SiO_2 + 2C + Cl_2 = SiCl_2 + C_2O_2$$

If we lead Cl over Si ~~by~~ at ~~ordinary~~ moderately high temps. a compd is formed containing the elements — 2 $SiCl_2$ HCl.

Note

A fluid — if evaporated to dryness — it inflames — With H_2O it is decomposed & there results 2 SiO + H_2O

This Compd resembles in its properties to Organic Compds. i.e. Si_2HO_3 (Formic Acid with Si for Carbon) — It gives, too, very similar products of substitution + decomposition as this Organic Acid — i.e.

$Si_2HO_3 + 3HCl = Si_2HCl_3$ — corresponds to a Cloride of Formyl. From these compounds the analogy of Silicon to Carbon in its behavior in what are called Organic Compds is perfectly established.

Fl_2Si.

Formed by ~~treating~~ a mixture of SiO_2 with $CaFl$, with conc. SO_3.

~~2 Ca Fl~~) (Ca O SO_3 / Ca O SO_3) 2HO

Si O_2) Si Fl 2

~~2HO SO_3~~

749.

It is a colorless gas - which smokes in the air, + is instantly decomposed by HO i.e.

$SiFl_2 + 2HO = SiO_2 + H_2Fl_2$

If however an excess of $SiFl_2$ is present, Hydro-fluo-silicic acid - thus.

$SiFl_2$, HOHO, $SiFl_2$, $SiFl_2$ } SiO_2, $HFl, SiFl_2$, $HFl. SiFl_2$

$HFl\ SiFl_2$ This acid corresponds exactly to an Oxyacid in which the O is substituted for Fl.

Properties It is much used as a reagent in the laboratory - & is important in so far as it gives us an insoluble precip. with

Ex KO salts - + is used in the separation of BaO from SrO + CaO.

The future field of experiment & investigation into the nature & analogies of the Si Compounds is vast; now that the possibility of bringing about analogous Organic Compds with Carbon has been proven.

Many of the Compounds found in nature & known as the 'Minerals' may by a careful investigation prove to be naught but analogously constituted Compds to many well known Organic Compds. The discovery of Wöhler has thrown a new & interesting light upon the domain of Mineralogy, which (alas, for its individuality as a science), may in a few years be swallowed up in the capacious maw of

757

that progressive science Organic Chemistry.

Zirconium.

Closely allied to Silicium + Carbon comes next, + closes the list of the non-Metallic Elements. It occurs in nature in many minerals — but all of them are "rare". Among them may be noticed the following: —

Zircon — $ZrO_2 SiO_2$ — classed among the precious minerals. Called ordinarily Hyacinth. In Norway it forms a constituent of a Syenitic rock. called from that fact Zircon Syenite. Then in the minerals Ostranit, Melakon, Auerbachite, Katapleit, Trachyaphaltite. &c.

Zr. **Lecture 83rd** Zirconium

If we treat the mineral Zircon like we do the Silicates we separate the two substances in the form of ZrO_2 + SiO_2 Then mix the two with Carbon + lead Cl gas over ~~them~~ through a heated tube - There forms $ZrCl_2$ + $SiCl_2$ - the latter is volatile + distills over The $ZrCl_2$ is a solid substance + remains behind. It can be purified by washing with ClH - which dissolves up all traces of $SiCl_2$ &c. From this compd we can form all the compounds of Zr. + ~~the~~ metal.

Metallic Zr.

Man. by treating the salt KFl, $ZrFl_2$ with Alumin-

753

ium - it dissolves out - the Zr - It resembles Si, very closely. Is not attacked by the atmosphere - but - like Si - it is vigorously attacked by the strong bases KO HO or NaO HO, The Only Compd as yet Known with O. is -

ZrO_2 :

Resembles SiO_2 in appearance - apparently 2 modifications, One, an easily + the other, a difficulty -

Properties

soluble modification similar to SiO_2 which it in all respects closely resembles.

It possesses the property of uniting with an acid or a base - By treating it with Conc. SO_3 + evaporating to

dryness gives a soluble salt – this distinguishes it from SiO_2
Is very difficultly fusible & phosphoresces very strongly.

The Metals –

Metals

As remarked in the introduction these bodies do not divide themselves very sharply. If we should proceed to the bodies which possess the nearest resemblance we should find the Iron group; that which would come next in order. But as we gradually come to more & more distinctive features – it is best to begin with those bodies which possess these in the most characterized manner. & hence we begin with Na, K. &c

755

<u>Natrium</u> group - Na
<u>Kalium</u> group - K, Rb. Cs.
<u>Magnesium</u> " = Li, Mg.
<u>Calcium</u> " = Ca. Sr. Ba.
<u>Yttrium</u> " = Er. Tr. Y.
<u>Cerium</u> " = Ce, La. Di.
<u>Aluminium</u> " = Th. Al. Be.
<u>Manganese</u> " = Mn. Fe. Cr. Ur. Ni. Co. Zn. In. Th.
<u>Lead</u> " = Pb. Bi. Cu. Cd. Hg. Ag.
<u>Gold</u> " = Au. Pt. Pd. Rh. Ir. Ru. Os.
<u>Molybdanum</u> " = Wo. Mo.
<u>Tin</u> " = Sn, Ti. Ta. Ni. Va.

Classification of the Metals into Groups.

It is best to begin at the most strongly developed character bearing metal for the reason that we have given on the opposite page. And hence we begin our consideration of the metals with <u>Natrium</u>

First group = Na.

The first two groups + the first named metal of the third group (Li) are distinguished by their intense affinity for Oxygen the action of atmos. moisture Ox. them.

Again. When in contact with H_2O they energetically decompose it forming an Oxide + H ↑

Again. Form the strongest bases known to us.

again. They under no circumstances can form Acids.

again. They form few very few insoluble ~~substance~~ compounds.

again - When not united with an involatile acid, they are upon application of heat completely volatile - an important property + distinguishing.

General Characters of the Metals of the 2 first groups + of the metal Li

757

Natrium

Vorkommen | Comes most universally spread in nature, slight traces in the atmosphere – as different salts, it occurs abundantly in the waters of the seas.

Oligoklase | In the solid Earth – it forms a constituent of many silicates – none however so abundantly spread as Oligoklase.

$NaONO_5$ – | Again (in Chili) in great beds as Soda Saltpetre

$3NaFl, Al_2Fl_3$ | Then as 'Cryolith' – in Greenland it occurs in vast quantities (as $3NaFl, Al_2F_3$) + from it the most of the Aluminium is manufactured.

NaCl | Lastly + most important, it occurs in vast beds – (particularly in the trias) as Rock Salt (NaCl) of universal use + importance – sometimes in the neighborhood of salt lakes – a product of evaporation.

It appears ~~that~~ this salt NaCl is necessary for the maintenance of animals & plants – for it is necessary for their healthful growth &c. — *In Organic World.*

At the beginning of this Century Sir H. Davy, by Electrical decomposition was enabled to Isolate the metal from the Oxide NaO. — *Darstellung*

In the ~~manufactory~~ it is produced by the hundred Pounds. by a simple reduction of the Carbonate with Carbon – generally $CaO\,CO_2$ is added – (as a separating agent) the mixture is given above –

C. 1.00	
$NaCO_2$ 3.64	
$CaO\,CO_2$ 0.68.	

$$NaO\,CO_2 + 2C + (CaO\,CO_2) = Na + 3CO + CaO\,CO_2$$

It is carried on in large Iron retorts + the Na distills into Naphtha.

759

It is allowed to run its course beneath Petroleum, on acct of its intense oxidizibility.

Properties | Is a silver white metal - of brilliant metallic glance. Sp. Grav. 0.97 - & hence swims upon H_2O.

Possesses a strong affinity to almost every other body (except strong metals) - Thrown upon water - it slowly decomposes

Ex | it ~~forming~~ NaO & H. but ~~without~~ inflaming.

If a thick gum is mixed with the Water - preventing the moving of the metal & thus confining the decomposition

Ex | to one spot - so much heat will be generated as to inflame - It then burns with the characteristic Yellow flame.

Owing to its intense affinity for Oxygen - Chlorine & all the Metalloids it is of the greatest importance to the chemist as a reducing agent - It will take Oxygen from all the heavy metals & reduce them to the metallic state - So also will K, Rb. Cs. Li Mg ~~Ca~~ Sr. Ba.

Use as a Reducing agent

$NaO + NaO_2$ (?)

The first is the only one well known, & ~~that is~~ universally known - Formed by direct combustion of Na in air - It appears to be infusible; is a greyish white solid, having an intense affinity for Water - which makes it an admirable substance for dessicating.

NaO

761

NaO CO_2 in Manufacturing ***NaO HO***

NaO CO_2 → an important salt - & from it we obtain the substance NaO HO - We simply mix a concentrated solution of NaO CO_2 with Caustic Lime -

NaO CO_2 + CaO, HO = NaO HO + CaO CO_2

this NaO HO can be evaporated in a silver dish & can be obtained crystalline - it is a real chemical Compd - of 1 atom of NaO with 1 of HO → is not ~~very~~ deliquescent, & exposed for ~~a~~ ~~a short time~~ to the air will not attract moisture ~~&~~ ~~enough to pass into the liquid state~~. It is much used in every branch of the Arts & manufactures.

KO CO_2 is the deliquescent salt, as distinguished from NaO CO_2

$NaO\,CO_2$

Is the most important of all the metallic Compounds. It is now universally obtained now from the salt $NaO\,SO_3$ viz:

$NaCl + HOSO_3 = NaO\,SO_3 + HCl$

then - $NaO\,SO_3$ / $C\ C$ } CO_2 / CO_2

$CaO\,CO_2$ } NaS / $CaO\,CO_2$

~~then~~ ~~NaS~~ / ~~$CaO\,CO_2$~~ } $NaO\,CO_2$ / CaS

Le Blanc's Method

Note

The first product is Hydro Cloric acid. The second reaction shows the reduction of the $NaO\,SO_3$ to NaS. & the third - the formation of $NaO\,CO_2$ by mutual interchange. By treating with HO - the $NaO\,CO_2$ is drawn out. & ~~NaS is~~ CaS left behind, & Sulphur Obt. from it.

763

$NaOCO_2$ Crystallizes with 10aq.
Easily soluble in H_2O — reacts strongly alkaline — by heating it gives up only slight traces of its CO_2.

With this H_2O which it crystallizes, it is never left combined but is freed from it by Heat — otherwise for every 100 lbs of $NaOCO_2$ we should be obliged to pack + send away 100 lbs of water.

Lecture 84th

From the Carbonate all the ~~salts~~ of NaO can easily be obtained. NaO SO$_3$ - Glauber's Salt occurs in many mineral springs. Crystallizes with 10 aq. By heating it looses its HO entirely. There is a Bisulphate of NaO; which as it does not loose its SO$_3$ except at a red heat - is used to ~~treat~~ substances with that acid at a high temperature, in many analytical investigations. NaO NO$_5$ (Soda Saltpetre) occurs in Chile & other places in great beds - is used in the preparation of KO NO$_5$ - & to produce other salts of NaO - being very hygroscopic it cannot be used as KO NO$_5$ -

Use of NaO HO 2SO$_3$ in Analysis

NaO NO$_5$

765

There are ClO_5 - IO_5 - BrO_5 - PO_3 - &c salts of NaO

Borax Borax - is probably the most important salt next to $NaOCO_2$, used for Solder & for beads of Metallic Oxides

The Compds of Na with Cl & Br & I. → The first, NaCl, occurs in great beds in several formations particularly in Trias. It is obtained. Either broken out &

Rock Salt sold as such. & is then used in the manufacture of Soda Then:- by evaporating NaCl Waters - (see Onondaga Salt Works). Possesses the property of being almost equally

Note soluble in cold as in hot Water - a behavior exceptional to ordinary salts.

Again - by evaporation from sea water - It is partially impure - Crystallizes in the regular system (∞O∞) - it melts + upon higher heating it volatilizes itself -

NaI - similar to NaCl

NaBr " " "

NaFl, a white powder - easily soluble in H_2O - etches glass.

Na + S.

If we reduce the salt $NaSO_3$ with H gas - we obtain Na_2S - (ditto by reducing with C) - is a yellowish powder - soluble in H_2O.

Besides this salt - we can produce other Na + S compd's - by melting Na_2S with S - Na_2S_5 is the highest sulphid.

767

Reagents for Metallic Oxides.

It is of importance to know these many metals by their reactions, just as we did the acids.

The reagents used for the separation of the metals are somewhat different from those of the acids. The method of separation depends upon the different behavior of the Sulphides Chlorides + Carbonates of the various metal groups. For this purpose we first get the metal or its oxide or salt, into solution — then add HCl to obtain the Chloride. Whether we obtain a precipitate or not — what is its color + solubility &c — of course will be remarked

Then we obtain the sulphide + for this purpose we have two reagents an acid + a basic sulphide (i.e. HS + NH_4S) - + its behaviour in relation to these is noted - color of precipitate, solubility in acids or bases. Then, the - <u>Method</u>
Hydrated oxide may be formed - by NH_4O,HO - or finally. the Carbonate by means of NH_4OCO_2 or $NaOCO_2$ Solution - + its behaviour noted - If none of these reagents produce the necessary Effect, viz:- of enabling us to distinguish them - then we must seek for special reactions - viz! - note the color of the flame the bead of Borax - or special precipitations must be sought

Ditto

Some of the metals do not give any particular reactions with these reagents - & ~~for this for~~ on this account we must have some other & particular reactions for these substances - Such a metallic

Reactions for NaO

oxide is NaO. it possesses only one or two insoluble compounds, & they cannot be used to detect or separate NaO from other metallic Oxides. So that we must have special & not general reactions for its detection.

The precipitates with SbO_5 & SbO_2 - are not characteristic enough to detect it in the presence of other metals — & the only reaction ~~&~~ left is that of the flame.

Reactions for NaO

HS.	NH₄S.	NH₄O HO	NH₄OCO₂
⊖	⊖	⊖	⊖
NaO HO.	NaOCO₂	2NaO HO PO₅	KCy Fe₂Cy₃
⊖	⊖	⊖	⊖

Special reactions →

SbO_5 ~~solutions~~ give, when brought into contact with fused NaO or its salts a precipitate insoluble in HO, (ditto SbO_2).

The best, + only reliable reaction – is that of the flame + the spectrum, It (i.e. NaO or its salts) gives a monocromatic (yellow) flame – so intense + pure – that all objects seen by its light appear yellow. (so paper stained red by HgI). When the vapor – which gives this flame – is examined through the spectroscope it appears as a bright yellow line. proving that the light emitted by the glowing vapor of Sodium is monocromatic

Group 2nd {Potassium, K / Rubidium, Rb. / Caesium, Cs.

The first of these metals is as widely spread as natrium - Occurring - as - Oxide always - & in combination with CO_2 - NO_3 - SO_3 &c & particularly with SiO_2 - forming a constituent of many of the most widely spread rocks - (Feldspar - Mica, Porphyry &c) i.e. Granite Gneiss. Mica Schist - Lavas. Basaltic & Porphyritic rocks. The latter two - are rare though wide spread, & were discovered by means of the Spectrum

Lecture 85th

Kalium

These first four substances (Na K. Cs Rb.) possess great resemblance, the second is not so abundant as Na. + the last two - are rare. **Occurrence** As Orthoklas - it is widely spread - from this rock it is depleted with - + finds its way into mineral springs &c. as KO CO_2 &c. **Springs**

KCl beds are too abundant + are very important - + are used to manufacture the metal - It is taken up by plants - as KO SO_3 + KCl - into the substance - Hence - the importance of a granitic - + basaltic soil for their growth + fruitfulness **KCl beds** **In Soil**

773

~~Tar~~ ~~Crate~~ of Potassa — form
ed in the manufacture of
wine by fermentation — is
an important material
for the manufacture — of KO.
Like all the Organic
Salts we have only to heat
to destroy the acid — & then
dissolve out the KO — from
the relics of C.

K — is manufactured like
Na — by mixing the Carbonate
Manufacture — with Carbon + glowing in
Iron ~~Cha~~ retorts — + distilling
over in a hydro-carbon,
It is a brilliant metal
of silvery lustre — soft +
wax like — Melting Point at
Properties. — 62°C. Specific gravity = 0,865.
Equiv, 39.2.

It is lighter than HO. When thrown upon it, it inflames & decomposes, the HO - forming $KO + H$ - (from $K + 2HO$), in inflaming it - burns with a brilliant violet flame, which is very characteristic for Potassium. Ex.

When heated to vaporization it is converted into a gas of a green color - while Natrium under similar condition gives a colorless gas. It possesses an even greater affinity for the metalloids than Na - as may be judged from its energetic decomposition of HO - with evolution of light. It is used very universally as a reducing agent - & will even take metalloids partly from comb. with Na.

Color of the Vapor of K.

Reducing Power

775

K + O –

Only One well known <u>KO</u>
Then there is a suboxide K_2O?
" " " " Superoxide – KO_2?

<u>Potassa</u>

KO.

Obtained just as NaO is obtained – Can be obtained anhydrous – brought into contact with HO – it instantly heats itself, & takes up another atom of HO.

<u>Manuf</u>

Again by treating $KOCO_2$ with Caustic Lime – it is obtained best :–

~~$KO\,CO_2$~~ ~~$CaO\,HO$~~ aq } $KoHO$ eaq. $CaOCO_2$ ↓

A decomposition analogous to that of the manufacture of NaOHO,

It is a white powder & crystallizes in thin lamina. It has Proper-
a great affinity for HO & for ties
CO_2 — from the air — Is soluble
in water to great extent. & reacts most energetically alkaline. the smallest trace
will give to a great HO mass
an alkaline reaction.
Being a very strong base, it is
used in analytical operations Use as
to precipitate metallic oxides a rea-
by extracting their acids gent.
from them — $PbOA + KOHO = PbOHO\downarrow + KOA$.
From $KOHO$ — all the other
salts can be manufactured.
$KOCO_2$ — obtained from the
ashes of plants — by crystallizing
out the other salts — $KOCO_2$ remains behind — & can be crystallized out with 2 atoms of HO.

$KOCO_2$ — As thus obtained it is in large beautiful crystals, & unlike $NaOCO_2$ – it is deliquescent seizing moisture with avidity from the air. & forming a liquid. As with $NaOCO_2$ we can form an acid Carbonate, $\left.\begin{matrix}KO\\HO\end{matrix}\right\}C_2O_4$ – by leading CO_2 through a solution of $KOCO_2$ – it is a stable & non-deliquescent salt loosing its (1 atom) CO_2 upon Heating

Properties.

Bi Carbonate of Potassa.

$KOSO_3$ – can be best obtained by decomposing KCl with SO_3 – & glowing – is Anhydrous & crystallizes in the Rhombic System. With H. or C. can be reduced to KS – Can be combined with another atom of SO_3 vs – $\left.\begin{matrix}KO\\HO\end{matrix}\right\}SO_3$

$KOSO_3$

Formerly used to manufacture SO_3. If we divide a portion of SO_3 HO, into two equal parts, neutralize one part exactly with KO HO, + add the other part to it — we shall have formed the salt, $\left.\begin{matrix}KO\\HO\end{matrix}\right\} 2SO_3$ or what is the same $\left.\begin{matrix}KO\\HO\end{matrix}\right\} S_2O_6$ this acid like CO_2 being bi-atomic.

General Method of forming Neutral + acid Salts

A most important salt — comes in nature — but always impure with KCl + NaCl, + is purified from them by boiling + crystallizing them out — from the slight quantity of NaCl &c remaining it can be freed by washing out with a solution of pure $KONO_5$ — which dissolves out NaCl &c but leaves the Insol. Saltpetre pure behind. It is much used in the arts — eg: — for preserving meats &c from decomp. on acct of its antiseptic properties

Saltpetre

Purification

Again, it is used in analytical operations as a vigorous oxidizing agent.

Gunpowder — The greatest use of $KONO_3$ is however in the gunpowder manufacture. The French powder has the following comp.

$KO\ NO_5$ — } KS — (KS_2 ?)
S. } (SO_2 ?)
C_3 } CO_2

The decomposition is not simple as imagined — for $KO\ SO_3$ — $KO\ SO_2$ — $NH_4O\ CO_2$ — KS a small quantity — & other substances are found in the solid residue of a powder explosion — the gases are CO — CO_2 — SO_2 — The action depends, of course, upon the formation, very quickly, of these gaseous bodies, which

Action of Gun Powder

The action of gunpowder is greatly influenced by graining; the more coarsely grained, the less is the real working power of the powder - because the gases find themselves under a smaller pressure.

Influence of the graining of Powder.

$KO\,ClO_5$ - is an important salt much used in oxidizing processes in the laboratory -

$KO\,AsO_5$ (acid + neutral) is a beautifully crystallizing salt -

$KO\,SiO_2$ - cannot be obtained crystallized.

Note

In newer times - this salt has been applied to a very important use - as a glazing namely - to glaze the hangings + scenery of theatres, which as it does not allow a planting of combustion process etc. [illegible]

Use of $KO\,SiO_2$ as a glaze + prevent. of Conflagrat.

781.

KCl

KCl – a beautifully crystallizing salt – KBr. KI – can be obtained in the usual way. All crystallize anhydrous, & are isomorphous – Crystallize in the regular System ($\infty O \infty$) – NH_4Cl. NaCl KCl – KI, KBr &c &c are isomorphous), These salts dissolved in H_2O – produce Cold.

KS & KS_5

KS – by the reduction of $KOSO_3$ by H gas or by C. It is much used as a reagent in dissolving up the acid Sulphides, of As Sb. Sn &c. in the separation of these from Pb. Ag &c. KS_5 – is most important – & is

Hepar Sulphuris

formed by smelting together $KOCO_2$ + Excess of sulphur – & heating to red heat. (Liver of Sulphur)

Lecture 86th

HCl	, H_2S.	NH_4S	. NH_4O.	NaO HO
⊖	⊖	⊖	⊖	⊖

Reactions

$NH_4O\,CO_2$	. NaO CO_2	. $\left.\begin{matrix}2NaO\\HO\end{matrix}\right\}PO_5$	+c +c
⊖	⊖	⊖	

Special Reactions → The first consists of the color of the flame which is of a beautiful violet color, (For these flame reac- tions - it is always best to make use of the Clorides, which are among the most volatile) - this can easily be used to distinguish between Na & K. When Na & K occur together : the flame of Na, is too intense to show that of K, so that the latter is invis- ible, by passing - however - the light of the flame through

Special Tests.

Ex

a piece or prism of Cobalt glass - or - an indigo solution, the yellow light of sodium is retained + the violet light only of Potassium is left through, this is the universally adopted method of detecting them when together -

2nd Reaction. Is unnecessary when we only wish to detect the presence of K. It is - the precipitation from HCl ~~xe~~ solutions by means of $PtCl_2$ when the double salt KCl, $PtCl_2$ is formed.

KO_2? KO_4?

A yellow solid → A higher oxide is formed by the combustion of Kalium in dry oxygen, It siezes moisture with avidity - + with HO - gives off O - + forms KOHO. composition not accurately determined.

Rubidium + Caesium.

Very rare - but wide spread. Found in Silicate Rocks - In Lepidolith, (Li. Ru. Cs). In many minerals waters in traces - again in the mother liquid of Dürkheim waters - again traces of Rb wherever K salts occur in great quantities - It is best obtained from the mother liquid of mineral H_2O's - We can separate K. from Rb. Cs in the following manner. The Mother-liquid concentrated is treated with $SnCl_2$ when there is precipitated out $(Rb(Cs)Cl - SnCl_2)$ impurified with Potassium - this precipitate is treated with NH_4OCO_2 + evaporated to dryness

Bunsen's method of separating Cs + Rb from K.

186

SnO_2 is separated + RbCl is separated –
Then the RbCl is again precipitated with $SnCl_2$ + this process is repeated three or four times at least until every trace of Potassium is removed – that can be removed by this method.
This RbCl gives a precipitate of (RbCl $PtCl_2$) in the cold. is boiled with water in – small portions – by this process the KCl $PtCl_2$ salt is dissolved out – while (Rb (Cs)Cl, $PtCl_2$) which is far more insoluble remains behind.
This is now Reduced in a stream of H + Pt is formed + RbCl, + CsCl separated.

The precipitation with $PtCl_2$ is again made - again boiled with HCl - & reduced & this process repeated till the spectrum fails to give the Potassium lines - then it can be regarded as free from all traces of K.

Rb & Cs must be separated by the different behavior of the bi-tartrates of these two bodies. The bi-tartrate of CsO is very easily soluble - & that of RbO difficultly soluble.

This salt of CsO + RbO is formed & treated with HCl - filtered - the filtrate - ~~reprecipitated~~ evaporated (i.e. evaporated to dryness) again treated with HCl & the process repeated until the spectrum shows the separation to be complete.

Separation of Rb & Cs. from Each Other Bunsen

787

General Method of forming Salts

$$\left.\begin{matrix} RbCl \\ HO\,SO_3 \end{matrix}\right\} \frac{RbOSO_3}{HCl} \nearrow \text{salt}$$

From the chloride or any other, we can form any desired compd. So the Sulphate from the Chloride.

$$\left.\begin{matrix} RbOSO_3 \\ BaOHO \end{matrix}\right\} \frac{RbO,HO}{BaOSO_3}$$

From the Sulphate, we can obtain the Hydrated oxide by carefully & exactly precipitating it with BaOHO.

$$\left.\begin{matrix} RbOHO \\ CO_2 \end{matrix}\right\} \begin{matrix} RbOCO_2 \\ HO \end{matrix}$$

From RbO.HO, for example we can, naturally - by neutralization manufacture - any combination desired. the carbonate - with CO_2 - the Sulphate - with SO_3 - the chloride, with HCl &c. the general method is given.

The tartrates have the following composition $\left.\begin{matrix}RbO\\HO\end{matrix}\right\} C_8H_4O_{10}$. So with CsO.

To obtain the metals we proceed as with Potassium the following is the mixture adopted by Bunsen. — Manufac. of metallic Rb &c

Carbon - 1 part
$RbOCO_2$ = 7.92 "
$CaOCO_2$ = 0.68 "
} also for Caesium.

Rb - is a silver white metal like Potassium - has a higher sp. grav. 1.52. & will not swim upon H_2O, is softer - & smelts - at 38.5°. Brought into the air it smelts. Atomic Weight = 85.36. Has a stronger affinity for Oxygen than even Potassium, & will reduce it partially from its Combinations, Decomposes Water even more Energetically. — Properties of Rb

789.

$RbO\,HO$ – like $KO\,HO$, deliquescent,

$RbO\,Cl_2$ – analagous to Potash –

$RbO\,SO_3$ " " $KO\,SO_3$

Salts of Rb. All the salts of RbO are isomorphous to those of Kalium

Cs.

→ Atomic Wght = 133.06.

Has not yet been obtained metallic, the ~~other~~ salts are precisely like those of K + Rb.

Cs + its salts &c in ~~chemical~~ physical character – & only by means of chemical reagents separable.

Solubility of the Rb. K & Cs Salts 100 pts HO dissolve at 0° –

$KO\,NO_5$ – = 13.3 | $KO\,SO_3$ = 7
$RbO\,NO_5$ – = 20.0 | $RbO\,SO_3$ = 32
$CsO\,NO_5$ – = 9.0 | $CsO\,SO_3$ = 160

The different solubility of various salts gives us the universal mode of separation.

+90

H S. { NH_4S { NH_4OCO_2 }, NH_4O – &c

for RbO. + CsO-salts – The flame, is impossible to be distinguish from that of Potassium, in all three cases it is an intense violet one. the spectrum alone, gives us the means to distinguish them from another. the method however, will form the toothpick(?) of another lecture.

Detec by Spectral Analiss.

Magnesium Group

Li } Lithium forms a
Mg } transition metal from the second group to the third.

It occurs somewhat widely spread – but in small quantities. Spodumen. Lepidolith Triphan &c. from these minerals – it is washed out

Lithium Occurrence

791

In Mineral Waters. by acid waters, & thus in a soluble form exists in the mineral waters of many springs particularly in those of the before mentioned Dürkheim Best obtained from the mother liquid of the mineral waters of Dürkheim.

Manuf. active. If we precipitate all the Cal Bal &c. by $NH_4O\,CO_2$ + evaporate till ~~dryness~~ all NaCl is crystallizes out; we shall have left behind, a mother liquid contain-ing Clorides of Li. K. Rb & Cs.

LiCl is ~~not~~ soluble in ~~[illegible]~~ $C_4H_6O_2$. By treating these with abso-lute $C_4H_6O_2$ - the LiCl easily dissolves out while the KCl, RbCl & C are very diffi-cultly soluble. By repeating the process the Lithium salt may be obtained quite pure.

In the ordinary way now - the other salts may be obtained. In one particular - too - the Lithium salts differ from those of the previously mentioned alkalies, viz: in concentrated solutions they are precipitated by a solution of NH_4OCO_2 (but only from conc. solutions), The metal & its salts forms a sort of transition ~~from~~ the strongly defined alkaline metals K. Cs, Rb + Na - to the metal magnesium, + both Li + Mg - form the transition to the strongly defined alkaline Earths - Ca. Sr. Ba.

Lecture 87th

Properties of Metallic Lithium.

The metal can only be obtained by Electrolysis, from the Cloride. It is a silver white metal, & it is the lightest of all solid bodies. sp. grav, 0.5 it will swim upon even Petroleum, It will decompose H_2O at ordinary temperatures as powerfully as K & Na.

Etc

It burns with an intense red flame in air.

Is an interesting metal in that it illustrates the fact the the lightest solid body should belong to the metals to which the heaviest, belong.

$LiOCO_2$

By glowing $LiOCO_2$ will not give up its CO_2 - like KO, NaO, &c, we can obtain the Hydrated-oxide, but from the sulphate.

$LiO\,CO_2 + KO\,SO_3 = LiO\,SO_3 + HO\,CO_2$

$LiO\,SO_3 + BaO\,HO =$

This Hydrated Oxide is somewhat difficultly soluble. When impurities KO or NaO are present it dissolves very readily. It is a peculiar fact that when pure many of the lithium salts dissolve difficultly — but when impurified with Na or K salts they become very readily soluble, this can be made a test of their purity.

Influence of impurities upon Solubility

The solution of $LiO\,CO_2$ reacts strongly alkaline, is not readily soluble & is stable in the air. When pure (absence of $NaO\,CO_2$) it can be precipitated by $NH_4O\,CO_2$. Perfectly volatile in the flame.

Many of the salts ($LiO\,SO_3$) crystallize beautifully.

Lithium + Magnesium in solution together — can be separated by $\left.\begin{matrix}2\,NaO\\HO\end{matrix}\right\}PO_5$ — when MgO is (in NH_4O solution) thoroughly precipitated — + can be filtered from the LiO solution. If this be now evaporated to dryness — + digested with little HO — the alkalies dissolve + the salt $\left.\begin{matrix}2\,LiO\\NH_4O\end{matrix}\right\}PO_5$ — remains behind.

Separation of Li from the Mg, Salts,

If the alkalies K Na &c be present with Lithium — we have only to convert them into Chlorides (after getting rid of CaO &c) + then ~~dig~~ treat with absolute $C_4H_6O_2$ — the alkaline Chlorides are very difficultly soluble — the LiCl on the contrary dissolves out

From the alkalies

For Lithium as with KO + NaO — the best reaction is the appearance of its glowing vapor. when passed through a prism This must be done — whenever the presence of Na or hide traces of Lithium. — Best Reaction the Spectrum

Lithium Salts give some precipitates — but all are soluble in NH_4Cl — like MgO: these — Insoluble Salt of Lithium

HS	NH_4S.	NH_4OCO_2	$2NaO,HO\}PO_5$ + etc
— o	— o	— o	NH_4Cl → o

Only with Phosphate of NaO, if we evaporate to dryness — + then add HO. it remains $2LiO,HO\}PO_5$. thus it can be separated from MgO. — Note

The great reaction, is however the flame: its salts give us a beautiful Carmine Red flame, When NaO Salts — Flame test

797

are present they hide the flame of Li, just as they do that of KO - hence whenever the two occur together we must use the spectral analysis, to separate or detect them.

Magnesium.

Very widely spread in nature - There are few silicate rocks which do not find some Mg salts. Then from these crystalline rocks it is dissolved out in various forms - Cloride, Sulphate &c - into Mineral Waters, &c,

Occurrence of Mg Salt in Nature

It occurs abundantly as Talk - $4MgO\ 5SiO_2$. Then as ($MgO\ CO_2$. Magnesit) - Then as (Dolomite $CaO, MgO\ C_2O_4$) forms whole mountain masses. Then as $MgCl$ or $MgOSO_3$ in many mineral waters - particularly the former.

If we smelt MgCl with Natrium - we obtain the metal. MgCl + Na}= Mg + NaCl. It is a white metal - like Ag - is stable in dry air - is crystalline - can be drawn out very fine into wires. Burns in the air with an intensely white flame - & exerts an immense chemical effect. It can be used to reduce the compounds of the metals which follow. Sp. Grav =

forms only ~~only~~ one Oxide.

Behaviour of Mg Salts &c

Salts of Mg.

MgO HO: like CaO HO - MgO - is very little soluble in HO: by heating it becomes less & less insoluble in Acids

MgO HO - little affinity for HO

+ is very <u>little</u> soluble in HO - by glowing it becomes less + less soluble.

$MgOCO_2$ - in Mineral Waters
<u>$MgOCO_2$</u> like $CaOCO_2$, dissolved in CO_2 - by precipitation with an alkaline Carbonate. Crystallizes with 5 HO. Looses its CO_2 very easily. by even boiling with HO, part of it is reduced
Magnesia to MgOHO; thus 'magnesia
<u>alba</u> alba' is formed. ($3\,MgOCO_2$ MgOHO). By glowing it gives MgOHO.

$MgOSO_3$ - important - Crystallizes with 7 HO. Called Bitter
$MgOSO_3$ Salt (Epsom Salts) - a medicine very much used. By glowing it looses its HO - + still more heating drives off part of SO_3

This $MgO\,SO_3$ can form double salts with $KO\,SO_3$ $RbO\,(CsO)\,SO_3$ &c &c – with 6 atoms HO, which have played an important part in the history of isomorphy

Double salts of $MgO\,SO_3$ with $KO\,SO_3$ &c &c

$\left.\begin{array}{l}KO\\MgO\end{array}\right\}S_2O_6 + 6HO$ a whole row of such compd. can be formed.

A very important salt of MgO is that of the double Phosphate of MgO NH_4O, the salt by which we separate MgO from other metallic Oxides – It has the property of being utterly insoluble in NH_4O Solutions –

$\left.\begin{array}{l}2\,MgO\\NH_4O\end{array}\right\}PO_5$ → gives the composition of this most important salt.

By glowing it is converted to $2\,MgO\,PO_5$ –

MgCl — MgCl crystallizes with 6 HO - very deliquescent + can be used as a desiccator.

By evaporating this salt to dryness there is formed partially MgO + HCl.

note — The anhydrous MgCl must be formed by glowing the double salt MgCl, NH_4Cl. When the NH_4Cl volatilizes + leaves behind MgCl.

The separation of MgO salts from those of the CaO group — Like Zn salts - the Mg ditto, form soluble double salts with NH_4Cl - + can thus be separated from the salts of CaO group - for when the double salt (MgCl + NH_4Cl) is treated with an alkaline carbonate - it (i.e. the Mg) is not precipitable, while the CaO group are under all circumstances thrown down,

Lecture 88th

Group third { Calcium = Ca, Barium = Ba, Strontium = Sr.

The Cloride of Mg. is very lightly fusible, & is adapted to the manufacture of the metal. Magnesia waters – (i. e those Containing MgCl – by evaporation & distillation give not pure HO but HCl. with the HO. → MgCl HO = MgO + HCl. (See the inconvenience to Steam ships) . Showed we attempt to crystallize out a mixture of NaCl + MgO SO3 – it would depend entirely upon phys-ical Conditions, – as to whether two or 4 salts would crystal-lize out. When they & other similar salts occur in min-eral waters it is impossible to say how they are Combined

$8O_3$

Reaction for Mg.

Presents the greatest analogy with Li. If we test for it without the presence of NH_4Cl - it has an entire analogy - except alone the precipitate with $\left.\begin{matrix}2NaO\\HO\end{matrix}\right\}PO_5$ in pres. of NH_4O = + this reaction alone enables us to separate the two Bodies. It precipitates out as $2MgO, NH_4O, PO_5 + 10HO$? In high temperatures the Mg. salts, unlike those of the previous ones of K. Na. &c are involatile. + hence for this substance we obtain no flame reaction. The chief reaction being the precipitation from NH_4O solutions with $(2NaO, HO)PO_5$

Calcium — Ca.

The group to which Ca belongs have the greatest chemical analogy to one another. Of these Ca occurs most widely spread in nature — in silicate Rocks — in almost all of them — In vast deposits as $CaOCO_2$ — in the form of the various limestones — Marbles — Chalk — &c &c — Occurrence From these original rocks & the Chalks &c, it is washed out by Carbonic acid waters & are in this form given to the plants, & in this way Carried out into the sea to supply the organic life there Corals, Conchifers &c) (Remark. The organic origin of sedimentary deposits of $CaOCO_2$).

805

Again, it occurs as CaO. $SO_3 + 2HO$ in various formations - mostly Tertiary - + Quaternary formations - Then as Anhydrite ~~(or Karstenit)~~ And as Dolomite ~~(as the CaO)~~; from these salts ~~it can be~~ the Oxide (CaO) can be obtained by solution (of $CaOSO_3$) in HCl or (with $CaOCO_2$) in HCl, HO -

Ca

Separation from other Salts

+ precipitation as $CaOCO_2$ with an alkaline Carbonate, by glowing this ~~at~~ purified Carbonate - there will remain behind CaO -

Manuf. of the Metal

The metal can only be obtained by Electrolysis - + then only with the greatest difficulty. It is only that in later times it has succeeded in obtaining the pure metal.

It forms a brilliant Brass yellow metal, of an intense affinity for O — will energetically — decompose HO at ordinary temperatures — Can be drawn out into Wire &c Must be kept in dr Hydro Carbons. Sp. grav = 1.566 - 1.584 — Properties

Ca + Oxygen

There exists a CaO_2? but is unimportant —

But the most important Oxide is the Ordinary CaO — (Burned Lime) — Can be obtained by glowing the Carbonate — — Ex

A white (or grayish — powder, which, when moistened with water — does not combine for some minutes — then the combination takes place suddenly, the whole mass heats itself, + a hydrate — CaO, HO is formed. — Properties of CaO + its Hydrate.

807

Uses. Its use as an ingredient of mortar, & for laying on walls in the form of a mixture of lad, H_2O + much H_2O, is as old as the culture of man. The Oxide (CaO) is generally manufac. from shells &c, mixed with Carbon & burned.

Properties. The oxide (CaO) is non vol-atile. Can unite with anhy-drous SO_3 with the phenom-enon of light. In solution

Ex. in H_2O it reacts Alkaline In such solution - it eager-ly attracts CO_2 from the Air. + forms an insoluble $CaOCO_2$ - With H_2O its ~~H₂O~~ Comp. is not stable but it can be driven from the CaO by red heat - while the alkaline metals will not loose it.

$CaO\,CO_2$ —

Occurrence & Uses.

Occurs in Nature very pure as Calcite & as Marble — & as Travertine — , Again, but more or less mixed with Organic & other impurities as Limestones of various kinds — marls, &c; all of which except the Calcite & Travertine have been deposited in beds — by the waters of the sea, from the debris of the shells & buildings of Mollusks & Coral animals. Marble is much used, for architectural & artistic purposes, & is exceedingly well adapted for the purpose. Calcite is used in many physical experiments in Optics, to illustrate double refraction &c, under the name of Iceland Spar.

809.

Ditto — $CaO\,CO_2$ is dimorphous, & occurs as Arragonite & Calcite one Rhombic & the other Hexagonal crystalline. The first is deposited from hot waters holding $CaO\,CO_2$ in solution, while the latter is deposited from cold solutions.

$CaO,\bar{O}$ — Oxalate of Lime $CaO,\bar{O}$, is an important salt, an insoluble salt, in which form we generally precipitate the CaO & by glowing convert it to CaO (or $CaO\,CO_2$).

Gypsum — occurs in nature very pure — in company often of $CaOSO_3$ + NaCl, it

Uses. — is much used as a manure — & the pure varieties, as a substitute for Marble in statuary — the most important use, however is ~~that~~ as a moulder in plastic work – the $CaO\,SO_3 \cdot 2HO$ in nature – possesses the property of loosing its water upon heating it – & upon moistening again to take it up – & becoming solid. thus adapting it beautifully to this use – by so doing it increases its bulk (note the result). The artificial Gypsum will not take up its water & loose it again – for the property depends upon the pressure – to which the other has been subjected.

Property of moulding dependent on pressure.

Increase in bulk.

Forms the cheif ingredient of the bones, white insoluble powder – Much used in Agriculture – & in the Manufacture of PO_5 & of Phosphorous

$3CaO,PO_5$ –

Occurs quite pure in Nature as Wollastonite – a simple salt of the composition indicated by the formula – $CaO\,SiO_2$

$CaO\,SiO_2$

811.

Ca Cl + its Properties.

A very important salt for the Chemist is CaCl, in the presence of moisture it is, upon evaporation very partially decomposes like MgCl (i.e. into CaO + HCl) It is a very deliquescent substance which speedily runs to a liquid in the air. + is used in many chemical + physical researches as a dessicating agent by ... in H2O Analyses, drying of gases &c.

8th

Lecture 89th

For gas analyses we dare not use smelted CaCl – for some CaO + HCl is formed which former might form a chemical union with the gas. **Note**

$\underline{CaS}$ – can be manufactured by reducing $\underline{CaO\,SO_3}$ with H gas or C. this must be carried on perfectly free from HO: otherwise there will result a decomposition →

$$\left.\begin{matrix} CaS \\ HO \end{matrix}\right\} \begin{matrix} CaO \\ HS \end{matrix}$$?

CaO HO is used ordinarily as a morter from its property of attracting CO_2 from air. Cement

(A mixture of HO containing CaO + MgO Silicates :) Ifine

813.

Manufacture + Properties.

fuse CaOHO + MgO HO, with SiO_2 - we obtain a Hydrated silicate of CaO + MgO - these possess the property of Hardening under Water.

Glass

Then Glass is a silicate of CaO + an alkaline Silicate, we distinguish various kinds of glass - some of which are adapted to chemical uses - + others not,

Kinds

($CaO\,SiO_2 + KO\,SiO_2$) (is Bohemian Glass) - it is easily distinguished by its being Colorless - again it offers the greatest resistance to all chemical reagents - + lastly it is very difficultly fusible which makes it adimirably adapted to Chemical uses,

Manufacture

<u>Soda glass</u> – shows a greenish color – & again, easily NaO attacked by reagents & <u>glass</u> lastly very easily fusible than 'Green Glass' – is a very impure glass – containing besides CaO KO & NaO; likewise Mn_2O_3 – Fe_2O_3 – &c. &c. – & is distinguished by its being Colored, being difficultly fusible, but then becoming suddenly very thin fluid.

Other glasses – are Lead Glass. so-an. Crystal glass – it is PbO the clearest, whitest, glass – <u>Glass</u> it has great refractive power – it is however, very liable to be attacked by reactionary agents – is very easily fusible & becomes black, from

815

Separation of PbO - (a reduction) upon being heated in the glass blower's lamp.

Then, the Borax glass are much used to imitate the various kinds of precious stones - on account of the property of BoO_3 to dissolve up metallic oxides & give their colors so readily - it is a mixture of Silicates with Borates.

Borax Glass.

The three substances CaO, BaO & SrO have very similar reactions: they agree in being non-volatile.

Reactions for CaO.

HS.	NH_4S.	NH_4OCO_2 -	$\frac{2NaO}{HO}$ PO_5
0	0	White	White

The best reaction to detect small traces with certainty is as with the alkalies, the flame (reddish yellow) & the spectrum -

Spectrum

816.

In concentrated solutions - SO_3 will precipitate out $CaO\,SO_3$ but only in conc. ones. Oxalic acid - precip. $CaO,\bar{O}$ - insoluble in $\bar{A}$ - a general test - and it is converted by Heating into $CaO\,CO_2$ or still more highly heated to CaO.

Special Reactions Ex ?

Strontium

Occurs like CaO - in traces with $CaO\,CO_2$ detectible only through the spectral analysis, it is not by far so widely spread as CaO.

Strontiant

As Carbonate ($SrO\,CO_2$) it accompanies metallic substances in veins &c.

As Sulphate ($SrO\,SO_3$), it occurs in company often with $CaO\,SO_3$ + Sulphur - particularly in Tertiary formations.

Caelestine

SrO, can be obtained by heating the Nitrate (not the Carbonate)

Manufacture of the SrOH.

817

Separation from Ba + CaO

From CaO + BaO - it can best be separated by means of the different solubilities of the Clorides + Nitrates of these three ~~oxides~~ Metals in $C_4H_6O_2$ - + this method is generally pursued in separating them when they occur together (explain)

Metallic Sr

The Metal can only be separated by Electrolysis - it is of the same yellow color as Ca - It decomposes H_2O energetically at ordinary temperatures, Reduces other substances taking from them O - Cl. Br. S &c.

Properties

Sr + O

There are several Oxides the most important is SrO. Can be manufactured by intensely heating $SrOCO_2$ like CaO.

note

It is a grayish powder - having the greatest &c resemblance to CaO; mixed with HO it heats itself highly - & combines with it to form SrO HO, & which they do not loose even at red heat. Can combine with SO_3 rather's with evolution of light & heat - must be kept in perfectly dry atmosphere. — Proper-ties — Ees

Will only loose its CO_2 - by the the most intense heating & then only in traces - — $SrO\,CO_2$

$SrO\,NO_5 + 5HO$ - is soluble - Crystallizes beautifully - it is insoluble in $C_4H_6O_2$ - While $CaO\,NO_5$ is soluble i.e. a method of separation of the two substances. — $SrO\,NO_5$ — $SrO\,SO_3$

$SrO\,SO_3$ more difficultly soluble than $CaO\,SO_3$

819.

The reactions of SrO are
Reac- identical with the general
tions- rea~~gents~~ctions for CaO.

	HS.	NH_4S.	NH_4O, CO_2	2NaO HO} PO_5
Ex	0	0	White	White

Special Reactions.

The best mode of detection is the flame reaction, & the spectrum. It gives
Ex an intense Carmine red flame. Then $CaOSO_3$ will give a precipitate of $SrOSO_3$ in tolerably conc. solutions - in dilute solutions only after some time.

For qualitative detection the best test is the flame. When together (CaO + SrO) the Spectrum only can separate them like NaO + KO salts when together.

The nitrate is used in preparing Pyrotechnical preparations - mixed with C? The partial volatilization of the Sr - staining the flame a beautiful crimson.

Uses in Feuerwerkerei, Ec.

Barium.

Comes more sparely in nature than SrO. In veins with metallic salts &c - as $BaO\,SO_3$ - & also as $BaO\,CO_2$. both being distinguished by their notable specific gravity, from the corresponding salts of CaO + SrO: from these it is separated as above mentioned by means of $C_4H_6O_2$. The BaO like SrO - does not give up its CO_2 - except in traces at intensely high temps - (manuf. BaO from $BaO\,NO_5$ -) as SrO was formed

Separation from CaO + SrO + the manuf. of the Oxide

82/

Ba & O.

BaO_2 = Super oxide of Barium
BaO. Oxide " "

The metal can only be obtained by Electrolysis but then not as regulus. will not smelt together — it is however, of a yellow Color — decomposes H_2O with the same energy as Ca & Sr

Metallic Ba

BaO

BaO — like SrO to verwechslung; gotten by glowing the salt $BaONO_5$ — it is a greyish powder — uniting with H_2O — with production of great heat to form $BaO.H_2O$.

Properties

Will unite with with SO_3 (anhyd) unter Feuerscheinung. Does not give up its H_2O even at red heat, like the others.

822

H.S.	NH_4 S.	NH_4OCO_2	$\left.\begin{matrix}2NaO\\HO\end{matrix}\right\}PO_5$
0	0	White	White

Reactions for BaO

Special Reactions.
Again, as with BaO, & SrO - the best reactions are the flame & the analysis of that flame in the spectral aparatus.

SO3 the best reagent in the wet way.

In the wet way the best reagent is SO_3 - which even in the most dilute solutions will bring about an utterly insoluble precipitate of $BaOSO_3$. ($CaOSO_3$ & $SrOSO_3$ will produce a precipitate of the Salts of Ba). The precipitate is insoluble in acids & decomposible only by fluxing with Carbonated

823

Alkali -

~~$BaO\,SO_3$~~
~~$NaO\,CO_2$~~

forms $BaO\,CO_2$ + $NaO\,SO_3$.

Again - a good reagent to separate Ba salts from those of Ca + Sr - is HFl, SiF_2 - which gives us an insoluble salt - particularly insoluble in alcoholic solutions in which it must be brought about,

BaCl } HCl + Ho
HFl Sif2 } (BaFl + $SiFl_2$)
+ Ho

↓

824

Lecture 90th

$BaOSO_3$, possesses the unpleasant property of attracting particles of other salts + holding them by contact action in solid state upon it

Unpleasant properties of $BaOSO_3$.

The nitrate + Clorate crystallize beautifully. The latter is much used in Pyrotechnical array. Mix the $BaOClO_5$ with 'milk sugar' + inflame with a glowing stick, it will burn with a brilliant green flame. BaCl is an important salt in analytical chemistry as a reagent for SO_3. + it serves likewise in preparing metallic Barium.

$BaOClO_5$ + milk Sugar

Exp

BaCl

BaO_2.

Barium

825

BaO_2 in a caked glass vessel & lead over it when, gently heated, a stream of dry & pure air. (wash with KOHO with SO_3 & with CaCl.

By dissolving it in a HCl we obtain HO_2, & BaCl, It can also be obtained in hydrated state. By highly heating it looses its one atom of O, & is reduced to BaO.

Electrolysis, It is now the proper time to explain many of the phenomena, which we have mentioned in the course of the consideration of the metals,

We will here explain the manufacture of the metals of the alkalies & alk. Earths by Electrolysis.

There are a number of elements whose affinities are so strong that, in the metallic condition they could not exist, but would instantly unite with other substances to form salts &c, such are the alkalies, & alk. Earths. We generally choose for Natrium the Cloride - Na the difficulty is that when the metal is formed it oxidizes itself instantly if it comes into contact with moisture, or anything containing Oxygen. If it comes into contact with the sides of the vessel it decomposes the glass - forming - NaO, SiO_2 - so that the utmost care must be taken - We make use of the following aparatus for its manufacture.

827

Manuf. of Na by Electrolysis

A graphite pole & an Iron one are placed in the smelted NaCl — the latter is now allowed to cool so far as to cover itself with a thin crust of Solid NaCl & the stream rapidly passed through, Cl separates at the graphite pole & Na — at the Iron one beneath the crust.

K & Rb

K. & Rb. only are reduced with the greatest difficulty

a Sub-Cloride is formed

It appears that a sub-cloride is formed — which is formed as the metal separates — hence the difficulty & length of the operation.

Mg

The use for magnesium — a mixture of the Clorides of K & Mg which has the advantage of being — fusible at a very gentle heat & is better adapted than pure MgCl. We must likewise constantly add NH_4Cl – for as we have shown MgCl in ~~contact~~ with moisture forms MgO + HCl – MgO is infusible Mg + prevents the Mg formed from collecting in a globule (for a class experiment, the arrangement in a clay pipe bowl will answer though much of the metal will be lost by oxidation.

For the pure manufacture – the vessel – as in the figure is used two poles of gas carbon are used – fitted into the lid (below) the negative pole – has step-like hollows which collect the Mg + prevent its rising to the surface & oxidizing – A partition separate the poles.

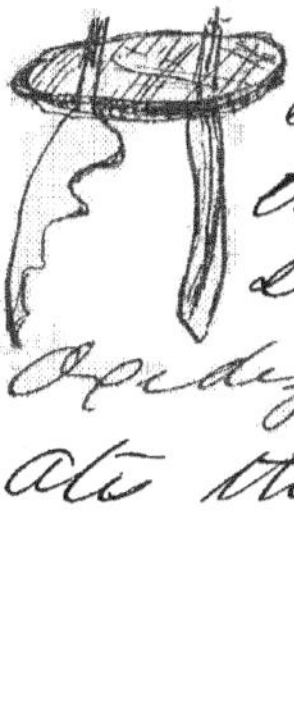

829

Lithium The reduction of Lithium is accomplished easily – In a tiegel of Porcellain – the negative pole is of gas carbon – the positive a needle of Iron, the pure LiCl is used – the positive pole must just touch the mixture – it collects in a few minutes + can be taken out with an Iron spoon.

El

For the darstellung of Ca, Sr, &c, the greatest difficulty is met.

Ca + Sr For Ca we use a mixture of CaCl + SrCl which fuses readily, + only allows Ca to be reduced.

Note → CaCl = 22 } Mixture for
SrCl 16 } manufac. of Ca.
NH₄Cl 1 }

The NH₄Cl must be added from time to time.

The use for its Manufacture an outer vessel of thon (porous) - about which is a metallic ring for the positive pole - this vessel contains the mixture, the inner ~~vessel~~ pole is like wire passed through a vessel of thon - the whole operation is attended with much difficulty + good fortune must attend the experim't.

Aparatus for manufacture of Ko.

Many metals can only be reduced from watery solutions. Always use the lowest compd's

In a H Cl solution the stream separates Cl at the negative pole - + H. at the positive - this H. acts to reduce the Metal - by secondary decomposition.

Reduction by H. of Cr

881

Advantage of having the H pole very small. If the surface upon which the H is set free be very large - a less amt of H is given off for a given amt of surface + the reducing power correspondingly small. If, however, we make the surface for the escape of H very small - a condensation of the amount of H at one point will be much greater + the reducing power - (for the same electrical intensity) very much greater than in the previous case. Hence in all these secondary reductions we make the H. pole as small as possible.

Lecture 91st

The light emitted is entirely independent of the chemical process or of any chemical changes going on at the time the light is emitted – but is produced at all times & under whatever circumstances the body may be made self-luminous.

To separate the various kinds of light from one another we pass the same through a prism, the various rays

Li (i.e. those of different rapidity, hence of dif- A Li ferent colors) have the proper- Light ty of being broken by a A Na prism with different light & degrees of angularity & so we can separate them.

833

A ray of red & yellow light emaning, for example, from A – would be refracted by a prism – at different angles upon the screen – & would appear separated – at R. & Y.

The Spectroscope. Is an aparatus by which the light from any glowing body is refracted or broken by a prism, & lead to the eye – this is accomplished by means of a tube with a very fine opening – A second ~~tube~~ placed in the proper direction leads the broken ray to the eye – A third tube – ~~contains~~ lights up a fine glass graduated scale – which is illuminated – by an ordinary gas lamp at its farther extremity

The following figure may (for one who has seen the apparatus before, & who, in addition, knew that the subject of Spectrum Analysis was under discussion), serve to call the various details of the instrument to mind; & the fact that the perpetrator don't pride himself on painting & sculpture (never having given this subject more than fleeting attention), may serve as an apology, to the connoisseur, whose artistic delicacy might be hurt by a contemplation of the figure in question.

835

Peculiarity of the light Emitted by Solid, liquid + Gaseous Bodies

If we notice now what light is given out by solid or liquid bodies we find ~~that~~ they give us ~~light~~ a mixture of all kinds of light, + when examined through a prism, we get for all solids + liquids a continuous spectrum.

If however we heat a gaseous body the case is very different. If we heat any gas to glowing (i.e. till it becomes self luminous) we do not get a mixture of all light vibrations but only one or two or more kinds (So for Na, for Li + c + c), showing that gaseous bodies possess only the property of giving out certain light rays

Na

Li + c

This law is strictly true for all gaseous bodies. & appears to be intimately connected with the nature of the body. & it is the most constant phenomenon that we can obtain, for any of the Elementary substances.

These lights examined through the split of the spectrum appear as lines in that part of the spectrum that corresponds to the color & these ~~lights~~ lines, too, always occur in the same place & by changes of temperature they become more intense & others are added to them

Spectrum lines for Elements

This property of the various glowing vapors & gases appear to be the most constant

837

No matter under what condition the substance may be found – no matter under what circumstances it may be volatilized these light it emits are the same, & the lines they form in the spectrum, are constant in place & number.

Depends upon the Chem. Charac. of bodies

SO that the rule may universally apply – that the light emitted is dependent upon Chem. Nature.

If we bring into the flame NaO PO_5 – NaO CO_2 – NaO HO NaCl &c &c – the same lines appears. Hence it would appear that compounds give us the same light as a simple one, but the rule cannot be universally applied.

Salts

If we illuminate the gas N - we obtain certain lines in the Spectroscope - C when volatilized C_2 gives us others - but Cyanogen - a compd - of C + of N. [illegible] (i.e. C_2N) - does not give us the lines of N + of C - but peculiar lines of its own. [illegible] So with some other Compds. Most [illegible] ~~Subs~~ compd's in an intense flame are [illegible] decomposed into their Elements (so with Na Compds + C) C_2N is not. [illegible]
To bring about such lights in the spectroscope we must choose the most volatile compounds of the substances used. + for all bodies it can be laid down as a rule - ~~that the~~ most adapted Compd. is the Cloride.

The lines are Reagents

It will appear from what has been said that the lines which a gas gives in the spectroscope are reagents — (we number the lines — calling the most distinct — α, the next β &c &c). & as the smallest traces, invisible to the eye — can

Sharpness

thus be detected, an idea may be formed of the value of this mode of detection — as a means of Qualitative analysis.

Extention of Analysis to Celestial bodies

It will further appear that, By this mode of analysis — it makes it possible for us to analyze substances which we have not directly beneath our hands — in this it differs from all other methods of analysis — for we have only to exam the light which comes from the body - to note the lines which it gives in the spectroscope - in order to come to a conclusion, as to its Constitution

Spectrum Analysis

We may examine any body which contains glowing substances ~~bodies~~ - (i.e. which are self luminous). No matter at what distance they may from us - & in this way in reality, the great fact has been demonstrated ~~that~~ we have a means of unfailing power, & of incalculable sharpness - to unravel at pleasure the mystery of the composition of worlds beyond our own. (So the comp- of the Sun. & of many stars)

Examination of lines &c.

841

Black lines — What is singular — is — that while earthly bodies, when volatilized give us bright lines on a dark ground, the heavenly ditto reverse the thing, & give us black lines on a parti-colored ground.

Explanation — Prof Kirschoff has explained the phenomenon as follows.

A certain definite relation exists between the absorptive & emissive power of a body at the same temperature — That is. When a body is perfectly black, which does not allow any light to pass through it — we call its absorptive power = 1. When ~~t~~ the absorptive power is ½ as large we call it = ½ &c. &c. Now — it has been discovered that when the <u>absorptive power</u> is large the <u>emissive power</u> is correspondingly large — when the former is small the latter correspondingly small.

Thus Elucidated →

		abs pow.	Emiss. pow	
[flame]		1/2	1	1
solid 1	[flame]	1/2	1	1½
solid 1/2	(			1/2
2	[flame]	1/2	1	1+1=2
6				6
6	[flame]	1/2	1	1+3=4

As the white light becomes more & more intense — the weak light of the gas becomes by comparison in the eye darker & darker — & when the former has become indefinitely increased ~~by~~ in intensity the light emit-

El 843 ted by the gas will appear black by comparison.

The fact can be demonstrated by actual experiment—

If we place a in curved light of great intensity—a salt of Na for example—& in a flame of small heat, a similar salt—the two side by side—will appear alike—but place the weak flame before the intense one—& it appears as a black spot upon it.

Lecture 92nd

solid		absorp.	emis.	
6				6 = 6
6	A	1/2	1	3+1 = 4
6				6 = 6 Exp.

gives us a black line — on a bright background so that we can at pleasure reverse the color of the lines.

Upon the sun's surface the conditions are such that the lines *must* be dark. The temperature of the sun as deduced from the amount of heat which can be condensed upon a given unit of surface — by a mirror — is so great that by it we can fuse Platinum &c &c — & it is a well known fact that —

Sun's Conde- *tion.*

the temperature at the point of emission, is far, incomparably greater than at the focus. So that, the temperat- upon its surface is so great as to volatilize all or nearly all the substances known to us - + hence these substances would ~~form a~~ ~~a metallic~~ Hydg. and metal. vapors. atmosphere - etc. + just as upon the earth HO is cont. in the air - + is deposited again in the form of rain - so there they may have their April showers of Platinum + Iron etc.

Inductive Reasoning upon the lines

The brilliant nucleus is within the gaseous medium without, + the black lines ~~are~~ of necessitated - Each one, which we can cover with a bright one.

We can thus carry our chemical
analyses to many millions of
miles into space - & discover
the composition of Suns & Stars.
With the planets it is hard-
ly necessary to say that they
give us a spectrum iden-
tical with that of the sun,
for they like the earth are
non-luminous, & glow by reflection.
With the fixed stars the matter is different, they are self-
luminous, & in all probabil-
ity like the sun, the independent centres of planetary Dist-
& solar systems, & the spec- ance
trum shows us different a disad-
lines from that of the vantage.
Sun - Their immense distance
(touch the argument upon light
its velocity & etc) is a disadvantage

847.

An examination has shown that many substances which are very rare upon the Earth – exist in abundance upon some fixed Stars f. Example – On Aldebaran & Tellurium is in abundance &c.

Conclusions upon Stars

Another examination gives the most interesting results with regard to nebulae – have been made – Some of these bodies are spherical in form, others irregular & covering vast celestial spaces. Some have been resolved into independent solar systems – complete & differentiated, others defy resolution with the mightiest telescopes, An examination of some of these nebulae

Nebulae

through the spectrum has given us the interesting fact that most of these irresolvable nebulae are nothing more than vast accumulations of Gases — for their spectrum gives us — bright bands in the places of Hydrogen & Nitrogen & another unknown gas — in a dark ground, a most remarkable fact — demonstrating their Chemical nature beyond a question. What the causes of their luminosity may be we are unable to determine unless indeed as Huggins — suggested — it may be caused by their motion through a resisting medium, as physicists declare the fact

Bright lines

They contain H & N & another substance

849.

A spectral examination during the late total eclipse of the sun, of the nature of the protuberances on the sun's surface – extending out far beyond it, & ending in rays, like the heads of saints are represented (Bunsen), have proven the said protuberances to be nothing more than immense masses of Hydrogen Gas.

Protuberances on the Sun – H

It is proven that H. N & Fe – are the most important materials in the constitution of the universe.

An interesting spectacle was offered to the scientific world some time since – in the conflagration of a sun, a small star appeared in an empty space

Note

& rapidly increased in brilliancy till it became of the 1st magnitude The spectrum gave a mixture of bright & dark lines — & **Note** that of H was particularly brilliant, It may possibly have been ~~that~~ a vast globe of H — & other gases, came into contact with, a globe of some elements for which it possessed strong chemical affinities — & ~~that~~ combustion at once set in — rendering the whole mass of gas & solid luminous from the intensity of the chemical action — It soon became less & less & finally disappeared entirely — this is not the first phenomenon of this kind recorded — Pliny

There is a method of Quantitatively determining

Volumetric Analysis. the quantity of any of the Metallic substances yet discussed.

We know that the Oxides of these metals are strong alkalies - an alkali has a peculiar effect upon vegetable colors - litmus is blued

Ex. &c. If then we to one of these solutions of Oxide - we should add an acid - (having previously added a ~~solu~~ tinct of litmus) - when the acid had exactly neutralized the alkali the coloring matter would just be on the transition point of changing - By taking an acid of known strength and adding a known quantity we ~~can~~ know the amt. of alkali which this quantity of acid can neutralize from the law of equivalents. We make use always of such solutions, that a cubic centimeter shall contain a weight of anhydrous acid or alkali equal to the Equiv. of said ditto expressed in grammes: such we call Normal solutions. By taking such solutions of acid we can analyze all the alkalies — by taking, vice versa — solutions of alkali we can analyze the acid. This process is called Alkalimetry + Acidimetry, or generally Volumetric Analysis. For the analyses, by weight

Eq

Normal Solutions

853

we can make use of the knowledge we have already obtained - i.e. that Ca. Ba + Sr Mg + Si form insoluble Carbonates - the alkalies insoluble double chlorides with Platinum.

Group V {Er. Y.
" VI {Ce - La. Di.

These substances belong to the most rare in nature - none of them being more abundant than is indicated by their occurrence in rare minerals.

Occurrence

These metals occur in several points of the Earth - in Norway + Sweden - + Finland - there too only rarely. They can only be separated from the substances already considered; for the whole group (in addition also to the Element Thorium), are precipitable by NH_4O - This precipitate can be dissolved up - strongly acidified + precipitated with Oxalic acid.

Separation from Ca + K, &c, group

These whole two groups have the property of giving with $KOSO_3$ double salts, some of which are difficultly Soluble - + by recrystallization they can be thouroughly purified from other substances.

Sep. of Y + Er group from Ce. Di + Ce

The double salts of the second group are insoluble in a strong solution of $KOSO_3$ - those of the Y + Er groups are ~~not~~ soluble - ! thus, we can separate the two groups from each other.

Lecture 93rd

Repetition of the Separation

Dissolve up - precip - with H_2S. filter + precipitate with NH_4O - dissolve in Oxalic acid - (dissolves Fe_2O_3. Al_2O_3, Be_2O_3) + these Earths Y &c remain behind as Oxalates - <u>Repeat to purify.</u>

By forming these double salts with $KO\,SO_3$ - + the double salts of Yt - + Er. are soluble in $KO\,SO_3$, the latter group are insoluble.

Group <u>V</u>. <u>Er. + Y.</u>

Separation of Er. from Y.

They occur in what is called <u>Gadolinit</u> (Norway) from this it is manufactured to the greatest advantage

Form the <u>nitrates</u> of these two Elements – (<u>the only way to separate them</u>) evaporate until NO_5 begins to decompose – a basic salt then forms itself $2\,YO; NO_5 + 2\,ErO\,NO_5$ – – the latter salt is difficultly soluble – the former is soluble – by recrystallization 8 or 10 times we have obtained Er pure as a nitrate.

[Margin note: Precip. as Ō salts slow. & treat – with NO_5 –]

[Margin note: Separ. of Y from Er & vice versa <u>Note</u>]

From this salt we can obtain the salts.

<u>Er.</u> Erbium

ErO is a solid substance of a delicate violet color

Is soluble in H_2O (?)

All salts show this color.

Er O O – thoroughly used.

857

Salts $3(ErO\,SO_3), 8HO$ – is isomorphous with the same salt of CdO + DiO.

$ErO\bar{A}$ – is beautifully crystallizable.

HS	NH_4O,	$NH_4O\,CO_2$	$^2NaO\, HO$ } PO_5 –
o	+	+	+

atomic Weight

Optical properties of Er Salts It has a peculiar optical property – by passing white light through a solution of a Er salt certain kinds of light are absorbed by the salt – + if we make a spectrum of the light passed through the solution the spec. will show us certain black bands – called absorption bands – (no resemblance to spectral lines in their nature

How to determine Atomic Weight

With the rare metals, of the V + VI th Groups – an in-direct method must be adopted for finding their Atomic Weights – namely – <u>indirectly</u>

We form as many salts as it is possible for us to form, until the crystalline forms, & the analysis tell us that we have formed salts isomorphous with some salt of known constitution. We then know if we have it to do with a prot oxide sesqui oxide &c, & by an analysis to percentages – we may easily determine how much of the base can unite with the Equivalent weight of the Acid; by subtracting 8 (O) from it we get the Equiv. of the Metal.

859

Yttrium.

Glow the mixed nitrates of YtO + ErO – strongly – the YtO NO_5 dissolves out easily – by recrystallization several times it can be obtained pure from the difficultly soluble $2ErO, NO_5$ –

IrO? Some chemists suppose Ir. to exist but their substance was always impurified with Er – (there must be no absorption bands).

This YtO shows much chemical analogy to ~~YttO~~ ErO – Is very soluble.

Salts are all colorless.

The sulphate is isomorphous to the ErO salt i.e. $3(YO\,SO_3) + 8HO$ – crystallizes in the same system & in similar form – so $3(CdO\,SO_3)9HO$,

This difference in the sol-ubility of the nitrates - is the only method of sepa-ration known to the chem-ist at present - other & better modes may hereafter be found - but at present do not exist.

Group VI {Ce. Di. La.

Occur in the mineral Orthit + in Cerit - both rare - occur in Norway. Pulverize the Cerit - mix with SO_3 + glow in a Clay dish - ~~[illegible]~~ + dissolve in H_2O. SiO_2 is left behind Lead HS through it on dissol-ving - to Precipitate Pb + cu. Then precipitate with NH_4O which will throw all these metals of the V, VI + VII groups

Sep. Mode of Procedure

861

Dissolve in Oxalic acid. (Fe_2O_3 Al_2O_3 Be_2O_3) 3HO) will be dissolved out

Then glow the residue + convert into double salts of $KOSO_3$ — filter + wash out with a concentrated solution of $KOSO_3$, in which CeO LaO + DiO are utterly insoluble — thus we have separated Ce La + di from the metals — Y + Er.

Separation

CeO has the power of forming 2 Oxides, the others only form one, + upon this fact the sep. is founded.

If we glow the oxalates with MgO, CeO is oxidized to Ce_2O_3 + there is formed a compound — $MgO\ Ce_2O_3$

If these we dissolve in NO_5 & we have

Separation of Ce from La & Di

Note

$C_2O_3 \; NO_5$

$LaO \; NO_5$

$DiO \; NO_5$

From this solution we can separate the Ce, quite pure, by treating with an acid – it has the property (from an treatment with an acid) of forming a basic salt which no other body possesses. Add a few drops of SO_3 – diluting & boiling warming – we get a precipitate of $3CeO,4SO_3+9HO$, almost entirely separated from La & Di.

Ce has been reduced by Natrium – it looks somewhat like Fe – is easily Oxidizable & must be kept under Petroleum.

863

Burns with a bright light. $3CeO, 4SO_3 + 9HO$, is the remarkable salt which we use to separate it from La & Di.

$CeO\,Ce_2O_3$ - an orange yellow powder - insoluble in all acids - except by evaporation with SO_3.

A mixture of the three oxides of Ce La & Di - is a orange color.

Ce_2O_3 has the property of forming a row of double salts (one of them is the following

$$\left.\begin{matrix} MgO \\ MgO \\ CeO \end{matrix}\right\} 3NO_5 + Ce_2O_3, 3NO_5)$$

most remarkable in their composition.

La. + Di.

The solution containing La + Di – is treated with Oxalic acid.
(Glow the Oxalate + convert into Sulphate – Heat to 150° – + dissolve in cold water – crystallize out – + the crystallized salts are insoluble)?...

Separation of La + Di

Dissolve in this cold HO – Heat to 40° or 50° – + suddenly LaO SO_3 separates out – + DiO SO_3 remains behind – repeat again + again + finally we get the LaO pure.
DiO – remains behind + is difficult to obtain pure; (no certainty about it as yet)

865

La, LaO.. + Salts

→ LaO CO_2 reacts alkaline?

La Salts Etc

Sulphate, is insol. when Hydrated — soluble anhydrous — + upon its sudden precipitation at a moderate temperature depends the separation of Lanthan. from Didym.

Lecture 94th

LaO SO_3 separates first - but it must be recrystallized at least 30 times it is doubtful whether it has ever been obtained pure.

Didym.

DiO the only oxide. (LaO colorless). The salts likewise violet - colorless.

$3(DiO\ SO_3) + 8HO$ is isomorphous with Er, Cd, &c salt.

~~Nw~~ Dy Cl - extensive violet,

Absorption Bands

Di - shows peculiar absorption lines - we need not pass the light through a prism - but only allow the sun's light to pass through the solution, upon a canvass the bands will be distinctly seen.

867

There are no special reactions as yet known for the separation of these groups V + VI; our whole process is to take advantage of the different solubilities of the different salts of these metals - + separate them - through the difference of their crystallization, possibly - the future will open better methods for their separation.

Reactions

Ce however has one peculiar reaction - by means of which it can be separated from every other body.

Special Reaction for Ce.

CeO dissolved in NO_5 + made as neutral as possible - will precipitate

MnCl solutions as MnO_2 a remarkable reaction, showing the precipitation of a Eq super oxide from a neutral sol.

$2 CeONO_5 + 3 MnCl HO = Ce_2Cl_3 + 3 MnO_2 (HCl)$?

Group VII
Al. Th. Be.

Thorium

Th is among the most rare of all bodies - comes in Orangit, a very rare mineral - in Thorit as a silicate of Thoria - it occurs very pure - we separate the SiO_2 in the usual way by evaporation to dryness with HCl - + dissolving up the base - the $Th_2O_3, 3HO$ is precipitable by NH_4O - separation from Al_2O_3 + Be_2O_3 by the behavior of its sulphate

869

ThO SO_3 at low temperatures is soluble - at about 100° - it is alterly insoluble - & upon this property, the possibility of separating this substance from the other rare metals depends. It separates as a white Crystalline solid: it occurs likewise in dilute solutions - which makes this mode of its separation a very certain one. The compounds yet formed: are →

Separation from Other Metals

ThO - a white powder.

ThO $\bar{O}$ - ditto.

ThO SO_3 + 5HO - the important salt by means of which it is separated.

ThO SO_3 KOSO_3 - likewise import.

We see that sharp reac-
tions, + definite differentiations
between these almost identical metal groups - entirely fail - + the method of
their separation may be regarded as a type of our
proceedure in all similar
cases.

Aluminium. Al

This substance occurs
in all eruptive & pluton-
ic rocks, + necessarily -
also in Metamorphic. It Occur-
found largely in that sense.
ingredient of these rock
called Feldspars. of
which there are many
kinds the most important
being Orthoklase, Oligoklase
+ albite.

871

These feldspars undergo a peculiar decomposition — in which the alkali in them is dissolved out by acid waters + a silicate of alumina — impurified with other bases is left behind called Kaoline, adapted to the fabrication of Earthen ware, Porcellain.

Kaolin

Cryolith, an import. mineral abundant in Greenland, used in the manufacture of metallic Al + of Na — has the composition — $2NaFl + Al_2Fl_3$.

Cryolith

Alum — is a technical product + a compound of Potassa with alumina — with SO_3 + HO — formula ($KO\cdot SO_3 + Al_2O_3\,3SO_3 + 24HO$)

Alum

Manufacture - as follows
The double salt, $NaFl, Al_2Fl_3$
(is converted into chloride?)
+ then reduced by Natrium (or used before)
The metal has a small
specific grav = 2.56 Is
very hard; is ductile - +
Malleable, into thin leaves
Is very stable - not attack-
able by dilute acids - but
attacked by alkaline so-
lutions. In the flame it
burns with a brilliant
flame, It leaves behind
a slight black mark on
paper - like Pb. It can be
used for household utensils
for - watchcases - Jewelry
Etc - but its uses for Household
Utensils is limited on acct of
its high price.

Ex

Uses

873

Al_2O_3

Al_2O_3

The only oxide which Al forms is Al_2O_3 — it is best obtained pure from the technical product called Alum — by dissolving in Water — precipitating the impurities with HS; reprecipitating the filtrate with NH_4O, & dissolving it in KOHO (to free from Fe), acidifying with HCl, & finally precipitating with NH_4O or $NH_4O\ CO_2$ — free from all admixture of foreign substances.

Manufacture

Properties

After glowing the Al_2O_3 is very difficultly soluble, & by smelting with KO_2H_2, S_2O_6 (or boiling with SO_3) it is rendered soluble.

This Al_2O_3 occurs very pure in nature as Corundum & as Ruby & Sapphire the latter two precious stones. Again, but impure - as Emery - much used as a powder for polishing metallic surfaces - grinding etc - owing to its extreme hardness (= Equals that of Corundum = 9)

Precipitable from neutral solutions by NH_4O, HO, best precipitated from solution as Chloride, as the Al_2O_3 has the property of forming basic compounds - ($ZnO.Al_2O_3$) Precipitated as a gelatinous white mass - insoluble in excess of NH_4O but soluble in excess of $KO\,HO$

Emery

Hydrate of Al_2O_3

Note

875

$Al_2O_3.3HO$ has a peculiar affinity for Organic coloring matters & is used as a permanent colorer (Lakes) for Linen &c. It forms with the coloring matter an insoluble compd

Lakes

Ep Bring into a litmus tincture — Some Al_2Cl_3 solution & throw down $Al_2O_3.3HO$, with NH_4O.

$Al_2O_3, 3SO_3+18HO$ | $Al_2O_3, 3HO$ dissolved in SO_3 gives us Sulphate of Alumina. used much in Coloring. Heating leaves behind Al_2O_3.

Alums. It forms an important row of double salts — Called Alums — much used in arts &c, in medicine, they are double Salts of Al_2O_3 with an alkali + SO_3 (with xnHO)

The general now may be considered as the foll. → Crystallographically important isomorphous all crystallizing in Octahedrons
($KO, SO_3 . Al_2O_3\ 3SO_3 + 24HO$) = KO Alum
($NH_4O\ SO_3, Al_2O_3 . 3SO_3 + 24HO$ = NH_4O Alum.
($NaO\ SO_3 - Al_2O_3\ 3SO_3 + 24HO$ = NaO Alum.
By heating these alums (except ammonia alum) lose their SO_3 + with KO is formed alum - $KO\ Al_2O_3$ - in which the Al_2O_3 plays the part of an acid.
Manufactured in various ways. Generally (En pros) - by treating Clays (pure) with conc. SO_3 + adding to the filtered solution of $Al_2O_3, 3SO_3$ so formed - a solution (or masses of $KO\ SO_3$ - Again from "Alaunite" - then from Alum-slate (Thon. B. Coal + FeS_2)

Manuf. Alum

~~877~~

This Alum-shist as it is Called is roasted in the air & the following results:—

Manufacture — $FeS_2 + 7O = FeO\,SO_3 + SO_3$ the SO_3, unites with the Al_2O_3 of the clay to $Al_2O_3, 3SO_3$) This is drawn out with HO,

Note (FeO SO_3 - is crystallized out —) $KOSO_3$ is added to the filtered mass, & boiled to ~~evap~~ conc— & finally crystallized.

All the soluble Al salts ~~are~~ give us an acid reaction - i.e. Color litmus paper - red.

The neutral salts all Contains - with one equivalent of Al_2O_3 - three Equivalents of - acid - O_3 - Al_2Cl_3 - $Al_2O_3, 3SO_3$ Etc —

Lecture 95th

The rare metals Y. Er. Ce, La, etc may be separated from the Metals of Group VIII — namely - Al, Be + Th - ~~from~~ the Clorides of the groups behaving differently - the Clorides of the latter group Al_2O_3 - Th_2Cl_3 &c - are volatile - those of Y - Er &c - are non - volatile. ?

Query - Can this separation be applied practically? RO unites with 1 atom SO_3 to form a neutral salt. Al_2O_3 however, always unites with 3 atoms of SO_3 (acid generally) to form the nearest neutral salt.

879

note → { $KO\ SO_3$
 $PbO_2 . 2SO_3$
 $Al_2O_3 . 3SO_3$ }

Three classes of bases - monacid - bi-acid & tri-acid bases The general rule may be deduced that

a neutral Salt. a neutral salt is one, which contains as many Equivalents of Acid in union with it - as there are Equivalents of ~~acid~~ Oxygens in the base - (It may nevertheless react acid).

We must form the Cloride by passing Clorine over a red hot mixture of

Al_2Cl_3 ← Al_2O_3 & Carbon - by heating the Al_2O_3 with HCl a basic salt will be formed - like the behaviour of MgO.

NaCl, Al_2Cl_3 - an important salt in the manufacture of Al.

By glowing with CoO NO_3 - any salt of Al_2O_3 - we obtain a beautiful blue colored compound - CoO, Al_2O_3? - is used as a test in the dry way - & more in the coloring art.

Test for Al

Ultra marine - formerly obtained from a natural production of that name & paid for with very high prices, used much as a coloring matter - Its manufacture by artificial means was at length accomplished.

Ultra Marine

Manufactured → by uniting NaO with SiO_2 - & mixing this compd with Al_2O_3 + adding Sulphur - & glowing - then by moistening with HO + glow

Manuf

ing again the beautiful blue color appears. Reactions for Al_2O_3 are very Characteristic.

Reactions	HS,	NH_4O,	NH_4S.	NH_4OCO_2	
	0	white	White	White	

note By NH_4OCO_2 it is precipitated as basic Carbonate, looses its CO_2 by simple heating. With NaO + KO a precipitate (white) ensues; but it is soluble in Excess of the precipitant — in which it plays the place of an acid.

test with $CoONO_5$ — With $CoONO_5$ —. the Oxide & Salts when glowed give a peculiar blue color — which is used as a test for Alumina Salts in the dry way.

There appear to be two mod- ifications of Al_2O_3 (allo- tropic) for when freshly precipitated by NH_4O, its hydrate is soluble partially Hence in precipitating Al_2O_3, 3HO, we always use NH_4S, which always contains free $NH_4O + HS$, in the latter - the Hydrate is utterly insoluble. Allotrop precip. by NH_4S. Note

The application of Clay - to manufacture of Clay- dishes - Earthen Ware, is traceable to the highest an- tiquity, The purest clay- is Kaolin - & is a compd- of $Al_2O_3, 2SiO_2$, it is white powdery - & furnishes the purest and best porcellain ware, - for; common wares an impurer Clay is used. Practical Uses of Those

813

containing besides Silicate of alumina - also $CaO \cdot CO_2$ SiO_2 - combined with Fe_2O_3 + with MnO + Mn_2O_3. as well as free in the form of fine sand. In HO it is plastic - by glowing it looses its HO, + becomes - hard - + (pure) impermeable - The various kinds are, Porcellain - Pipe clay, Capsule Clay + c + c,

Porcelain + c.

Beryllium

Be

Its Occurrence.

Comes in a precious stone - Beryll - Hexagonal crystallizing Crysoberyl - Aqua marine. On the whole a very rare substance.

It behaves very similarly to Al_2O_3 - so that when separating Al_2O_3 - we obtain Be_2O_3 in solution too.

Be_2O_3 however is soluble in $NH_4O\,CO_2$ – Al_2O_3 is precipitated. Note in Ex.

Be is a metal very analogous to Aluminium. Of small sp. grav. White & ductile. Metallic Be

Only one Oxide, (Be_2O_3) a white powder – the hydrate is like that of Al. The salts however are somewhat different from Al salts – Be_2O_3

Be_2O_3 gives no Alums. They have all a sweet ~~taste~~ (Glucina). $Be_2O_3\,3CO_2$ looses its CO_2 more difficultly than Al_2O_3.

Reactions Ex.

Reactions →

H_2S	NH_4O	NH_4S	$NH_4O\,CO_2$ →
~~O~~	white	white	white: Sol. in excess

886

The reaction with NH_4O-CO_2 serves us to separate Be_2O_3 from Al_2O_3. The salts of the first soluble in Excess of reagent - those of the latter are not.

Sep. from Al_2O_3

Most salts are poorly Crystallizing.

Group VIII. Mn. Fe, Cr. Ur Ni, Co. Zn. In. Tl.

As yet we have had only metals - precipitated by NH_4O as Hydrated Oxides & also by NH_4S as Oxides - (i.e. Al group + Ce. + Y group)

Propertes of the Group

We now come to a group of vast importance, which are distinguished by being precipitated as Sulphides by NH_4S → & in that reaction Characterised.

Manganese.

Is found everywhere - always in company with Iron Ores in traces - & abundantly spread in nature - & important - Occurs tolerably pure in many minerals - viz! — Pyrolusit - MnO_2 - Braunit - Mn_2O_3. Mnganit. Mn_2O_3, HO, these are the most widely spread of mangan. ores.

Mn

Occurrence

These generally contain traces of Fe_2O_3, sometimes CoO NiO, sometimes CaO, MgO, &c. By dissolving we get all these bodies in solution - To purify the Mn, from these impurities - we pro- Add $NaOCO_2$ to the solution & the ...

Separation of Mn

887

acid bases (Fe_2O_3) is pre-
Note → cipitated first - for - these
salts of sesqui-oxides -
are - all acid - thus re-
Ex moves $Fe_2O_3 3HO$ - in a yet
acid fluid - while MnO
+ CaO &c remain - by
Precipitation with NH_4S
we obtain it pure.

Reduc- The metal may be
tion of Obtained by reducing
the the Oxides - with Carbon,
Metal in an intense heat -
Or: - by reducing the
fluorids - by means of
Natrium - or: - by redu-
cing a mixture of MnCl
+ CaCl by means of Na-
trium - the first & second
are most practical.

Lecture 96th

It resembles Iron very much is capable of a high polish. & is stable in dry air - oxidizes in moist air like Fe. - non-ductile - This metal decomposes H_2O upon addition of an acid. Ordinary raw Iron is an alloy of Fe with some Mn.

Metallic Mn

Mangan. + O.

Oxides.

MnO = Protoxide of Mn.
Mn_2O_3 = Sesqui Oxide. "
MnO_2 = Binoxide " "
MnO_3 = Manganic acid
Mn_2O_7 = Per. manganic "

MnO = Protoxide of Mn

If we dissolve any oxide in HCl - & boil to solution we obtain MnCl - from which, by precipitation with

NaO,CO_2 + glowing ...

It is a greenish solid – stable in dry air – It is –

MnO a base – capable of uniting with acids to form Salts –

Upon exposure to air & moisture it becomes brown – by the formation of MnO, Mn_2O_3 – so with the higher oxides, if we ~~bad~~ glow them they give up O. + form these oxides.

Properties +CR

MnO – gives a whole row of salts – (mono-acid)

$MnOCO_2$ – a white powder gives up its CO_2 by heating.

MnO,SO_3 – isomorphous with $MgOSO_3 + 7HO$. with $KOSO_3$ $RbO(CsO)SO_3$ &c forms isomorphous double salts with 6HO (like $MgOSO_3$)

MnCl - dissolving any salt in HCl, smells very easily. easily soluble in H_2O, is stable in air.

Reactions are as follows,

HS.	NH_4O.	NH_4S	$NH_4O\,CO_2$
⊖	White.	flesh col.	White.

NaO - White - MnO HO,

By treating a prot-oxide salt of Mn - with a hypo-clorite (NaCl, + NaOClO) &c, we obtain a precipitate of the Super Oxide MnO_2 - Clethe with a little alkalie.

Reactions

Exp

The Borax oxidizing Pearl of Mn Compds - is Amethyst Pearl. Colored - reduction Pearl - colorless. The bead with NaO-CO_2 + $KONO_5$ - is a green colored one. (characteristic)

Exp

891

With NO_5 + PbO_2 => the Mn.

Reac-tions Ex: Salts color the liquid deep red - upon boiling - from the format. of Mn_2O_7.

In NH_4Cl solutions MnO Salts are soluble - (like

note: MgO Salts), a valuable property, enabling us to prevent the precipitation of MnO Salts by NH_4O - in separating them from those of Fe_2O_3, which are precipitated - (those of MnO remain in solution.).

Mn_2O_3

Manufactured by glowing MnO NO_5 (or MnO_2) for

Manufacture: some time - in combination with SO_3 it can be obtained by treating MnO_2 with Concent. SO_3

+ gentle warming. It is a good oxidizing agent.

By the gentlest heating the oxide is decomposed into MnO, Mn_2O_3. By adding HCl we get a black solution – by boiling – $MnCl$ is form $MnCl$ + Cl – It plays the part of an acid generally like Al_2O_3 (MgO Mn_2O_3).

Mn_2O_3

The know however of one salt of Mn_2O_3 viz – the Sulphat $Mn_2O_3 \cdot 3SO_3$. formed as reaction indicates

$$\left.\begin{array}{l} MnO_2 \\ MnO_2 \\ SO_3, SO_3, SO_3 \end{array}\right\} \begin{array}{l} Mn_2O_3\ 3SO_3\ aq \\ O \end{array}$$

By glowing this compound it is decomposed. The Oxide is not stable when heated but forms MnO, Mn_2O_3.

893

MnO_2 -

Occurs in Nature as Pyrolusit finely crystallized. A conductor of Electricity - is of much technical values in manufacturing, O, & Cl. By glowing in the air it forms MnO Mn_3O_4 is neither an acid nor a base - It is formed

Ex - in the laboratory, by treating any Salt of the Protoxide - with an Oxidizing agent. (NaO ClO, NaCl for Example) - It is a good oxidizing agent - by simple heating it gives off nearly all of its one atom of O forming MnO -

MnO_3 = Manganic acid.
formed by smelting together
MnO_2 + KOH, see – ! –

MnO_2
MnO_2 } KO MnO_3
MnO_2 } Mn_2O_3.
(KO) KO
~~HO~~ HO

MnO_3

This Salt is soluble in water ~~with~~ a rich Proper. green Color. (Character – ties – istic for all its Salts).
By standing in the air – the fluid soon changes its color passing rapidly: into blue – + red (Chameleon fluid)
$3 MnO_3 = Mn_2O_7 + MnO_2$.
~~The~~ Dropped upon paper. Etc the organic matter – rapidly seizes upon its oxygen – + the green color is changed to black (MnO_2)

Properties — By treating a solution of KO, MnO_3, &c. with SO_2 it is bleached to a colorless liquid + $MnO\,SO_3$

Ex — $KO\,SO_3$ is formed.

The salts of MnO_3 are isomorphous with SO_3 salts — but we only know of a few.

Mn_2O_7 — Salts are formed by leading Chlorine into a solution of

Manufacture — $KO\,MnO_3$ — the salts all possess a beautiful red Color. It can be separated for a short time. i.e. by adding very conc. SO_3 or PO_5 to such a solution. When the Mn_2O_7 separates & comes to the surface in the form of oily drops of a green color.

With SO_2 the salt $KO\,Mn_2O_7$ instantly bleached to a colorless fluid. $KO\,SO_3$ & $MnO\,SO_3$ being formed. Droped upon Paper — the rich red, is rapidly changed to black — (MnO_2 being formed. The salt KO, Mn_2O_7 is widely used in the volumetric analysis of Iron — the method will be discussed under the head of Fe.

With SO_2 Exe

Exe Paper

Use of Mn_2O_7

Mn S.

Formed by precipitation of MnO salts with NH_4S it is of a flesh color, when first precipitated — but is unstable easily oxidyzing — dissolves readily in HCl

$MnS + HCl = MnCl + HS$

MnS.

Exp

Iron

General Remarks

Is, by far, the most important of all the metals - the most widely spread in nature of all. - & having the greatest connection with the cultivation of Mankind. The geological history of man - teaches us to distinguish, first an age of Stone (implements being made from it); then an age of bronze & Cu) & lastly with the dawn of his cultivation the age of Iron, - & the civilization of a land can be accurately measured by its progress in the extraction, & the (abundance & utility) ~~[illegible]~~ of this ~~mine-~~ metal.

Occurs as Fe_2O_3 - Red Hematite - (Roth Eisen Erz) - seldom pure from Clay. + Silica - (Occurrence)
Then - $Fe_2O_3 + 1HO$, Brown Hematite - often mixed with more or less Clay. → again Fe_2O_3, $2HO$, Yellow Iron Ore
Thon Eisen stein - a mixture of Clay with Brown Hematite.
Then as $FeO\ CO_2$ - (from which the best steel is prepared) it occurs - + very abundantly in the coal period, + associated directly with Coal. It is the best (or one of the best) materials for the extraction of the metal as $FeO\ Fe_2O_3$ - in Magnetic Iron Ore - it forms whole mountain masses.

Lecture 97th

The metal is prepared in high furnaces. of one form - to be continuous, in the manufacture so that it is kept burning for years - the Fe - continuously removed below. Alternate layers of Fe_2O_3 &c - with Coal & Slag are placed in the oven & heated by - the bellows. The Slag is added for the purpose of allowing the metal to smelt together - otherwise it separates as a black metallic powder.

Manufacture

The Slag

We use for this purpose - a simple flux - of silica + CaO; if the ore contains the one, we add more

of the other - & vice versa -
The ore generally contains SiO_2
whence we generally add
$CaCO_2$ to it. as slag. (ordinary Limestone)
The slag smelts & carries
off the fine surface of
Oxide & other impurities,
& acts in every respect like
Borax - (the latter is too dear
for the purpose) - the metal
sinks to the bottom - is drawn
off continually as it collects.
& the slag does the same -
& can likewise be drawn
off when it collects in
inconvenient quantity.
By this means - pure Iron
never is obtained - for
the first product always
Contains about Six per
cent of Carbon - with more
or less Silica - S. & P.

Raw Iron 6% C

We may obtain two products by the first process viz. Raw Steel Iron, + Raw Iron. First is formed by adding little Coal + correspondingly much Iron Ore – we obtain – Raw Steel Iron. It is the hardest of all Iron – harder than steel + cannot be worked on this account – it contains Carbon in chemical combination, + does not separate on the surface as Graphite.

Raw Steel Iron

Raw Fe → Contains from 3 to 6 pr. ct. C. part chem. united, part mechanical. It is easily fusible – but is not adapted to hammering + forging – though from its fusibility it is of all kind best adapted to Casting.

Raw Iron

It is much softer than Raw Steel Iron, & is used for Casting &c &c - Can easily be fused, & is adapted to all the uses in the arts where the casting can be carried on - It cannot be ~~hardened~~ hammered. 92% pure Fe. Raw Iron

By puddling, (or refining) the raw Iron is robbed of most of its impurities; of much Carbon, SiO_2 - S & P. & Steel, or Bar Iron formed. Carried on by smelting the raw Fe - beneath a stratum of Coal & then passing through it a continuous stream of air by which the Fe becomes purer & purer, till it reaches 99 - or 80 %. Puddling Stable Iron. Bar Iron

903

The impurities are then burned off forming CO, + SiO_2 the latter entering into composition with some FeO or Fe_2O_3 + going into the slag.

	Roheisen	Stahl	Stab Eisen
Fe =	92.3	97.8	99.37
C =	3.	1.7	0.60
Si =	4.5	0.5	0.03
Cu. Mn. Pb. S. P =	0.2	0.0	0.00

This table gives the relative composition of the various kinds of prepared metallic Iron, beginning with the most impure.

Bar Iron.

This fe is the purest kind of metal which is manufac. It possesses the property of being ductile in the cold + when heated — does not smelt — but becomes soft can be hammered +

+ in every way is adapted to uses in the arts. whence it is called Wrought Iron Piano From this are prepared the <u>Strings</u> fine Piano strings - almost pure Iron, used for all purposes, Railroad Iron, axels &c. Can become crystalline in texture.

Steel -

Steel is a Carbide of Fe. of very homogeneous structure - Manufactured in very different ways.
Bessemer's process is the best, + most widely used.
We ~~[illegible]~~ heat the pig Iron (Raw Iron) till fused + allow the smelted mass to flow in great retorts - + from tubes air is forced into it. + at this tempe-

Bessemer's <u>Process</u>

ature the fe- burns. - the C + Si &c burn off when the process is stopped - (which can be done at stages where, either steel or later where Bar Iron is made) - the steel is run into moulds prepared for the purpose.

This steel allows itself to be worked in all ways. When it heated to a certain temperature, & suddenly Cooled - it has the property of becoming very hard & tenacious. This steel is no real chemical compound. more a mixture - By glowing & slowly cooling it becomes softer & can be adapted to other uses

906.

Has the property of retaining
magnetic force permanent- By
ey. &c. Iron can be obtained Electro-
perfectly pure by Electrical lysis
decomposition of the Clo-
ride. | Properties

Pure Fe is very brittle - &
is very soft - Can be filed
&c - is infusible - & is very
ductile. Sp. grav - 7.8439. ←
Is easily oxidizable - in
presence of air & moisture
is readily Oxidized in short
time - In a finely divided
state - thrown into the flame
it burns with a bright
flame. Iron will decom-
pose Water, at a red heat
but not at ordinary Decom-
temperatures - a fact with posing
distinction. (& with acids see) HO

$Fe + HOSO_3 = FeOSO_3 + H$

907

Fe + Oxygen.

FeO - prot - oxide of Fe

Fe_2O_3 = Sesqui " " "

Fe_3O_4 = protosesqui " " "

FeO_3 = Ferric acid?

FeO = Prot Oxide.

formed by heating together Fe_2O_3 with Fe powder.

FeO $\left.\begin{matrix} Fe_2O_3 \\ Fe \end{matrix}\right\}$ $3FeO$ It is a dark colored solid - according to Bunsen. according to ⇒ Gorup - Besanez it is not bek amt isolirt.

The most important salt is, $FeO\,SO_3 + 7HO$ - green vitriol - forms a whole row

Salts ⇒ of Isomorphous double salts like MnO & MgO

$KOSO_3$, $FeOSO_3 + 6HO$
$KOSO_3$, $MnOSO_3 + 6HO$
$KOSO_3$, $MgOSO_3 + 6HO$ } Rhombic

908

$FeOCO_2$ is an important Ore of Iron. best for steel. $FeOCO_2$

$FeCl$ — prepared by dissolving $FeOCO_2$ in HCl — when the air is excluded — (i.e. in CO_2 atmosp) $FeCl$

HS O	NH_4O white — (green to brown)	NH_4S black	NH_4OCO_2 white (soon brown.	Reactions

The changes are due to the formation of Fe_2O_3 salts (or Fe_2O_3 *per se*). Ex

With (KCy, FeCy) — it is first precipitated white — but almost immediately turns blue — from a mixture of some Prussian blue.

FeO Salts — like MgO + MnO — salts are rendered soluble to NH_4O &c by the presence of NH_4Cl — hence to test for Fe we always convert the protoxide into a sesqui-oxide salt — & perform the tests for them. Properties

909.

This is easily accomplished by treating any salt of the protoxide with an oxidizing agent (i.e. fuming NO_5 . $KOClO_5$ &c) — Ex

Fe_2O_3.

Formed by oxidizing by any salt of the protoxide with NO_5 + precipitating it with NaO, KO, NH_4O &c – thus — Manufacture

$Fe_2O_3, 3SO_3$ aq. } $3NH_4O\,SO_3$ →
$3NH_4O + HO$ } $Fe_2O_3, 3HO$ ↓ — Ex

It is a reddish powder infusible + stable — when glowed – it is difficultly soluble in acids. — Properties

It is very hard – used in polishing metals + mirrors – in a finely powdered state — Caput mortuum

910

Fe_2O_3 is a tri-acid base its neutral salts contain 3 equivalents of acid viz.
→ $Fe_2O_3\ 3SO_3$ ←

$Fe_2O_3\ 3HO$

$Fe_2O_3, 3HO$ - is easily obtained by precipitating a salt of the sesqui-Oxide by an alkalie - by heating, it looses its HO, + is converted into the difficultly soluble Fe_2O_3:
$Fe_2O_3, 3SO_3$ - a ~~brownish~~ white solid - crystallized it contains 9 atoms HO. Though it is soluble in all proportions in HO - it requires much time - sometimes days + weeks, to completely dissolve - like - RO_5 - see previous lectures - It forms an alum. ($Fe_2O_3 . 3SO_3 + KOSO_3 + 24 HO$)

Note
Forms an Alum

911

Lecture 98th

Salts of Fe_2O_3

Fe_2Cl_3 a yellow solid - formed by burning Fe in Cl gas, is somewhat hygroscopic. Behaves on evaporation like MgCl - viz forms some Fe_2O_3 + HCl.

HS	NH4O	NH4S	NH4OCO2
white (of S)	Red brn	black FeS	red → White

Note Reactions Etc

With HS - at first a blueish substance is precipitated - but soon disappears + white Sulphur is precipitated - + the salt is reduced to one of a protoxide - With NH4O - a brown-red hydrated Oxide is thrown down - insol. in Excess of reagent. NH4O CO2 - gives at first a reddish White precipitate which soon changes into a red one of $Fe_2O_3 3H_2O$. KCy-FeCy-> gives us the most delicate indication for the presence of Fe_2O_3. In the faintest traces - a blue precipitate of Ferro cyanide of Iron is thrown down. Sulpho Cyanide of K - likewise gives a beautiful red tint in the presence of Fe_2O_3 salts

Continued.

Etc

Fe_3O_4 =

Formed by glowing Fe, in the air - When the workmen heat Iron - & hammer it, fine - thin particle fly off - & are an impure Fe_3O_4 - It comes in nature in large Regular Octohedrons as Magnetic Iron Ore.

Occurrence

It may be regarded as a mixture of the two oxides $FeO + Fe_2O_3 = Fe_3O_4$ — & is generally so regarded.

FeO_3.

Ferric acid

Cannot be separated from its Compds. Is only known theoretically formed — If we suspend $Fe_2O_3, 3HO$, in Concentrated KOHO — & pass a stream of Chlorine gas through it. If we attempt to separate by means of an acid — Red Oxide of Iron is precipitated.

Ex

It forms with KOHO a violet red solution, which decomposes easily of itself — when allowed to stand for some time in contact with air.

Qualitatively Iron can be detected in the smallest traces by means of KCy-FeCy.

Quantitatively - it is always determined by means of titration with Mn_2O_7 in Comb. with KO. Called Chameleon Solution - from the Changes in its Color - in the titration of Iron Titrd - in its solution as FeO Salt. How The Compound $KO\,Mn_2O_7$ of Fe ~~when~~ is brought drop by drop into a solution of a proto salt of Iron - & each drop as it touches - Ecc the solution is rendered Colorless - till all the Iron is Oxidized to sesqui-Oxide - when the

915

first drop of Mn_2O_7 in excess – obtains, & gives its red color to the whole solution – so that the point of complete neutralization can be accurately determined.

Mn_2O_7	of complete neutralization can be accurately determined.
FeO A	$5 Fe_2O_3, 3SO_3 + A$
FeO A	
FeO A	$MnO SO_3$
FeO A	$MnO SO_3$
FeO A	
FeO A	
FeO A	
FeO A	
FeO A	
FeO A	
$HOSO_3$	
$HOSO_3$	

SO_3 HOH

The above reaction indicates the process of the analysis, showing the deoxidation of the Mn_2O_7 to $2MnO$ (or Mn_2O_2) + the oxidation of FeO to Fe_2O_3 salt.

As soon as every particle of 2FeO is Oxidized to Fe_2O_3 → the next drop of $KO\,Mn_2O_7$ will not be bleached — but will remain colored, so that the point of complete oxidization can be accurately determined

The Iron or in whatever condition of oxidation it may be in is dissolved to a protoxide salt — With Metallic Fe this is accomplished by dissolving in such a flask with valve — with SO_3 — in an atmosphere of CO_2 — a weighed portion of the Iron is of course used. The Sesqui-Oxide would be reduced with Zinc to protoxide

Method of Analysis

Solution

917

$Fe_2O_3 + Zn + 2HOSO_3 =$

$2FeO\,SO_3 + ZnO\,SO_3 + HO$ - the atom of liberated H - oxidizing itself at expense of Fe_2O_3.

Formula for determ

We need to have a normal solution of Mn_2O_7 salt (i.e. one whose strength we know) - with which to make our analyses - It is formed as follows

Ex

A quantity of Chem. pure Fe (piano wire) is diss. in SO_3. + placed in beaker - HO added - + some SO_3. Then from a Burette so many C.C. unknown Solut. Mn_2O_7 added till completely oxidized. We then know how many C.C. Solution = our known weight of Iron - + can calculate how much Iron - one C.C. Solution would equal - this value is all we need.

Cromium. Cr.

Occurs in ~~some~~ silicate rocks - Gabbro Serpentine - Then as Crome-Iron Ore. from which Cr + its Compounds are obtained **Occurrence**

Cr can be reduced from its Oxides by Carbon at the highest temperatures. It is, however only obtained pure by Electrolysis from its Cloride - CrCl.

Has much resemblance to Iron in Color - is however perfectly non-ductile being very brittle - is stable at ordinary temp's - burns at the highest temperatures Decomposes HO at white Heat + in presence of an acid - NO_5 will not attack it.

Crom + Oxygen

CrO = Prot- Oxide of Cr.

CrO, CrO_3 – Proto-Sesqui " " .

Cr_2O_3 = ~~Cromic Acid~~ Sesqui Oxide " " .

Cr_2O_3, CrO_3 = Cromate of Sesq. Ox of Cr.

CrO_3 = Cromic Acid.

Cr_2O_7 ? = Per Cromic " .

CrO_3

Is the most important Oxide – formed in combination by pouring over a mixture of KO, HO, + Cr_2O_3, a quantity of Conc. NO_5 + evaporating to dryness – by this means there is formed a Bi-Chromate of Potassa – a substance crystallizing in beautiful large crystals – of a bright red Color – formula $KO, 2CrO_3$

From this compd we obtain all the compds & the acid itself - By adding SO_3 to a solution of $KO,2CrO_3$ (SO_3 - in great excess.) - the acid CrO_3 - is separated - & CrO_3 gradually crystallizes in Deposites fine red crystals - It is then filtered (see later) - & the tra-ces of SO_3 washed off with Washing with NO_5 - NO_5 - heated to 75° to drive off NO_5. It is a beautiful red solid - Note crystallizing in large crys-tals - has great avidity for moisture - is a strong Oxidizing agent. So that if we place the acid upon paper - it is instantly (in sunlight) reduced to Cr_2O_3 Exp as is indicated by the color Changing from Red to Green.

921

Ex CrO_3 + $C_4H_5\text{-}OHO$. If CrO_3 is brought into contact with $C_4H_6O_2$, so violent a reaction ensues, that at times the $C_4H_6O_2$ will inflame - the CrO_3 is reduced to Cr_2O_3 which continues to glow. (contact phenom)

Ex CrO_3 + SO_2 Brought into contact with SO_2 Solution - the latter instantly deoxidizes it - under the formation of SO_3 + Cr_2O_3. HCl - decomposes it as follows, see →

$2CrO_3 + 6HCl = Cr_2Cl_3 + 6HO + 3Cl$

Reactions CrO_3 precip. $AgO\,NO_5$ - sol. brown as

HgO solutions Red as - Cromate of HgO. CrO_3

PbO solutions a beautiful Yellow ~~precipitate of~~ PbO CrO_3.

The salts are all colored. The neutral salts are isomorphous with SO_3 salts – The $PbO\,CrO_3$ – cannot be used in Coloring – though a very intense one, because of the property of Pb salts to be affected by H_2S; in course of time – even the infinitely small traces in the atmosphere will suffice to blacken the painting.

Why Crome Yellow is not unpainting.

$BaO\,CrO_3$ being insoluble & $SrO\,CrO_3$ soluble – the use of $KO\,2CrO_3$ as a reagent for separating BaO from SrO is often made.

Separation of BaO & SrO by CrO_3

Cr_2O_3 = Sesqui Oxide.

Easily obtained by reducing the CrO_3 by means of a reducing agent. or we can reduce the Cromates.

Manufacture

923

Manu- $Hg_2O\ CrO_3$, or $NH_4O\ CrO_3$ -
fact. when heated leave behind
Cr_2O_3 - when glowed it -
is exceedingly insoluble
in Acids - + is very stable
By leading a volatile
Cromate through a glowing tube - Cr_2O_3 crystallizes - + indeed it is isomorphous with Fe_2O_3 -
Rhombohedric - R = 86°
It is a remarkable
substance for contact
phenomena. Ipure

Ex. Heat $NH_4O\ CrO_3$ - it
commences to decompose
of itself, a combustion
with evolution of light
+ heat ensues. + Cr_2O_3 is
left behind.

Lecture 99th

Cr_2O_3 upon an alcohol lamp accompanied with a draft of air. produces the same phenomenon as platinum sponge - i.e. produce contact phenomenon of glowing continually. Exp

There are two allatropic modifications of Cr_2O_3. one green (ordinary) + the other blue.

Cr_2O_3 fuses with $KOSO_3$ + SO_3 an alum - > $KOSO_3, Cr_2O_3 . 3SO_3 + 24HO$ - Note in large beautiful large crystals (Octahedrons) - these of a dark color - can by putting them in a solution of ordinary alum be covered by a transparant crystal of latter. 2 Alums about each other

925

Alum Manufacture — Formed by treating a solution of $KO\ 2CrO_3$ with SO_3 + allowing to crystallize.

Cr_2Cl_3, can not be obtained by dissolving Cr_2O_3 in HCl + evaporating to dryness (while Cr_2O_3 or $Cr_2Cl_3 + Cr_2O_3$ is formed) (Cr_2Cl_2O)

Must be manufactured by passing over a heated mixture of Cr_2O_3 + C — a stream of Cl. — the salt is utterly insoluble in H_2O — but if the smallest trace of CrCl be added to it — it becomes soluble — a remarkable instance of the working of a contact substance

Merkwürdige Erscheinung

HS	NH_4O	NH_4S	$MnOCO_2$	Reac-
O	green Hyd	ditto	basic Carbon.	tions.

Borax gives in oxidizing + reducing flame - a beautiful green bead. (Characteristic) Ex

CrO_2

A mixture of $\left\{\begin{matrix} Cr_2O_3 \\ CrO_3 \end{matrix}\right. = 3CrO_2$

If we bring into $KO,2CrO_3$ solution HS - the reduced Cr_2O_3 combines with the CrO_3 still in the fluid. or by direct mixing of the two oxides,

Cr_2O_7

If we add to $KO,2CrO_3$ a quantity of HO_2 - a deep blue color ensues - If we pour upon it C_4H_5O it dissolves out the substance - it cannot be separated for it is exceedingly decomposible. Manufacture

927

CrO.

If we treat the compound Cr_2Cl_3 by passing a stream of H gas over it + we obtain behind – CrCl. It cannot be dissolved in HO, but decomposes the water + H is given off – the precipitates are brown-colored. The Oxide per se has not been yet obtained pure.

Manufacture

$Cr\{^{O_2}_{Cl}$

Cloro-chromic acid = $Cr\{^{O_2}_{Cl}$ Is a remarkable substance can be regarded as a Cromic acid – where Cl takes the place of one O.

$$3(Cr\{^{O_2}_{Cl} = {}^{CrO_3}_{CrO_3} + CrCl_3$$

Etc

Obt. by distilling a fused mixture of $KOCrO_3$ + NaCl with SO_3.

928.

We now return to the mode of filtering the CrO_3 formed in the previous lecture. As it must be washed with Conc. NO_5 to remove SO_3 - the ordinary mode would not do. So we use an air pump filterer + a glass tube - with a partition of artificial Pummice stone. It is then surrounded with a ring of Iron furnished with a lamp - + heated till perfectly dry + all NO_5 is driven off (not over 100°C). at the same time a stream of dry air drawn over it then brought instantly into a prepared dry vessel. as it is very hygroscopic.

Filtration of CrO_3

Ce

Uranium:

Sparcely scattered in nature as UrO, Ur_2O_3 = Uran Pech Erz - (same compos. as FeO Fe_2O_3) then Uran - glimmer but rare. From the first the metal + its salts are formed. We proceed as follows. We dissolve the ore in Aqua regia ~~dissolve~~ + evaporate with HCl several times. Pass through it a stream of HS, to remove traces of As + Pb which often accompany it, the FeO is oxidized with conc NO_5 + supersaturate the solution with NH_4O, CO_2 - in such a supersaturated solution - salts of Ur_2O_3 are not precipitated by NH_4S - we can then precipitate out the Fe + crystallize out the compound - $(NH_4O, Ur_2O_3), 2CO_2$; can be obtained by Electrolysis from UrCl - note

It is a very hard metal - resembling in color + properties nickel or cobalt. Is tolerably stable in the air, but can be smelted at a white heat, + then burns with brilliant flame, to Ur_2O_3. Properties

Decomposes HO - like all the metals of this group - in presence of an acid. or highly heated By NO_5 & HCl, Ur, is attacked - forming when evaporated - Ur_2O_3 - + Ur_2Cl_3

Ur + O.

UrO = prot-oxide of Ur.

UrO, Ur_2O_3 = proto-sesqui " " "

Ur_2O_3 = Sesqui Oxide " "

Any Comp'd or ox- glowed in the air leaves behind the UrO, Ur_2O_3 - behaving here just as we saw manganese behave.

Ur_2O_3.

By treating any salt of the protoxide with an oxidyzing agent - it cannot be precipitated by an alkalie as hydrated Oxide. To obtain it we must ~~glow~~ gently heating this compound — Ur_2O_3 NO_5. This last compound gives a number of absorption lines - (blue) - the crystals fluoresce -

Ur_2O_3 can play the part of a base or a weak acid, as a base, it, (unlike Al_2O_3 or Fe_2O_3 + others), forms neutral salts with one + not with three atoms of acid.

Ur_2O_3, an Exceptional Base

Ur_2O_3, AsO_5 -- ditto PO_5 -- is insoluble - hence the Ur_2O_3 salts are sometimes note used as reagents to analyze quantitatively AsO_5 + PO_5.
If we add to Ur_2O_3 solutions KOHO we do not get Ur_2O_3, 3HO, precipitated - but a compound of KO with Ur_2O_3 as an acid is formed - which is insoluble. Here Ur_2O_3 plays the part of an Acid.

933

Pearl + Reactions	The borax pearl is yellow in the oxidizing flame + green in the reducing flame – + at the same time the pearl ~~[illegible]~~ shows phosphorescence
Ex	HS & } NH_4O gelb $Ur_2O_3 . NH_4O$ } NH_4OCO_2 white – sol in Ex } NH_4S black
Note	$NaOCO_2$ gives a white precipitate – soluble in Excess.

With KCy, FeCy it gives a red-brown precipitate of UrCy–FeCy – a characteristic reaction – being the only substance – not precip. by HS – which gives a precipitate with such a color.

934.

Lecture 100th

UrO.

Manufacture

If we lead H gas over Ur_2O_3 - in the red heat we obtain this Oxide. The reduction does not go any farther than to the production of the Oxide (& not of the metal). **Properties**

This Oxide gives us ~~a~~ green salts - do not crystallise

$UrO\ Ur_2O_3$ = a compd - ~~bla~~ in which the Ur_2O_3 plays an acid part.

The same compd with Na base is $NaO\ 2\ Ur_2O_3$ → $(KO, 2\ Ur_2O_3)$ ← 2 atoms of the base uniting with 4 of the acid - or what is better One of base - to two of Ur_2O_3;

The subs. $UrO.Ur_2O_3$ is a black powder

935.)

Ni.

Chiefly occurs in Kupfernickel. Nickel Kies.

Occurrence

Used extensively in the arts: for alloys (German Silver) is in use in some countries as an ingredient of Coin.

Wherever Ni occurs so does also Cobalt. (As frequently) These ores are pulverised & are roasted with gradual heating to prevent melting. As ~~goes off~~ is converted to NiO AsO₃ + CoO, This is now heated with HCl by which the As compd is destroyed.

Then the FeO is treated with NaO ClO₅ + the oxide is precipitated out as Fe₂O₃ after the As.

see next page

The As is then precip-
itated with H_2S as As_2S_3
+ filtered off → then oxydized + Fe_2O_3 precip. removed
Then the CoO is precip-
itated next, by NaOCl
which precipitates the Co
first - as Co_2O_3. → this
is filtered off, and the Ni
is left alone behind +
is reduced by Coal.
Thus obtained it contains
½ to 1% of Co + As + c.
There are other methods
by means of which Ni
+ Co can be accurately sep.
The metal is easily re-
duced from its Oxide
by H_2 gas. The metal
has Sp. gr = 8.38. is difficultly fusible ~~+~~ like Fe.
is very ductile + is

937

stable - & is capable of taking on Magnetism like Iron; (is used for Argentine metal)

Dissolves in HCl & NO_5 — but best in Aqua Regia

Oxides.

NiO = Prot- Oxide

NiO, Ni_2O_3 = Protosesqui Oxide

Ni_2O_3 = Sesqui "

NiO_2? = Bin—> "

NiO.

NiO —> Is the chief oxide - obtained by glowing any volatile salt - of NiO - (NiO NO_5 &c) -

Salts —> An allotropic modification exists - by gently heating an alloy of Cu & Ni - this oxide crystallyzes out upon the surface - it is utterly insoluble — crystallizes in Reg. Octahedrons.

HS	NH_4S	NH_4O.	NH_4OCO_2
0	(black) NiS	Green — Sol. in Ex.	Green. Sol. in Ex.

If we treat $NiOHO$ — with Reactions ↑ $NaOClO$, $NaCl$ — it is converted into Ni_2O_3 — but slowly — Ex

Ni gives a pearl. in Ox. flame like Fe → in the reducing flame greyish from metal. Ni.

$NiOCO_2$ — green (apple). always basic. $NiOSO_3 + 7HO$ — crystallizes in Klinorhombic System. $NiOSO_3 + 6HO$ — crystallizes in the Quadratic Systems

$KOSO_3 + NiOSO_3 + 6HO$ — Rhombic
$NH_4OSO_3 + NiOSO_3 + 6HO$ "
RbO " " " " " Isomorphic
CsO " " " " " Salts

Being one of the group of oxides forming these isom. salts.

939.

$NiCy, KCy$ = Double cyanide of Ni + K. an impor. Salt from the behavior of which, we are enabled to separate Ni + Co. quantitatively

Ni_2O_3.

Obtained by Oxidizing the freshly precipitated, $NiOHO$. by $NaOClO, NaCl$.

Ni_2O_3 Does not play the part of a base — for we know of no Salts — is a black solid — dissolved in HCl gives off Cl — like Mn_2O_3 &c — + is converted into $NiCl$.

NiO, Ni_2O_3.

NiO, Ni_2O_3 By glowing NiO — at certain temperatures — this mixed oxide is formed. No important chem. facts are connected with it.

Co.

The greatest analogy to nickel chemically + physically. Wherever Ni occurs – there Co is found – mostly with As. → found as Cobalt glanz – Cobalt blüthe – Speiss cobalt (Co_2As?) &c. Occurrence

It is manufactured in the fabrication of nickel. Can easily be purified. (by H_2S + NH_4O) –

It has the same Sp gr = 8.538

Is white – of silvery lustre – is very difficultly fusible – Can be readily reduced from its oxides or its Chlorür by C. or more readily by H. Is magnetic – + does not loose its magnetism upon Heating, while Fe + Ni

941

loose it perfectly. Is very stable. Is readily oxidized at high temperatures—

Properties

Decomposes H_2O like all the metals of this group—at white heat, or in the presence of an acid (Co. Fe. Mn. Ni. &c)

Dissolves in HCl—in NO_3—HC—& gives salts of the Prot-oxide

Oxides.

CoO = Prot oxide

CoO, Co_2O_3 = P. Ses. "

Co_2O_3 " Sesqui "

Co_6O_{10} = Unknown?

It forms as will be seen a peculiar oxide—of singular & deviating Constitution.

The salts are all of a reddish color -

CoO. **Manuf. + Properties**

Formed by glowing the Hydrated Oxide in a stream of CO_2 - So also the Carbonate -

HS. { NH_4O } NH_4S { NH_4OCO_2 }
[illegible] {blau sol. ex. | black CoS | red - ex. sol. }

Borax gives a beautiful blue color in both flames

$NaO\ ClO$, $NaCl$ gives a black coloration + precipitation almost instantly. Ex in salts of the Prot Oxide.

Separation from Nickel

$CoO\ SO_3 + 7HO$ - Klino Rhombic

$CoO, SO_3 + 6HO$ - Quadratic -

$KO\,SO_3 + CoO\,SO_3 + 6HO$ - Rhombic.

$RbO\,SO_3 + CoO\,SO_3 + 6HO$ " &c.

$CsO\,SO_3 + CoO\,SO_3 + 6HO$ " Isomorphism

$NH_4O\,SO_3 + CoO\,SO_3 + 6HO$ "

943

Smalt — is a glass which is Colored by CoO. is much used as a coloring matter — when ground to an impalpable powder. — its use is now (except to the Chinese) superseeded by Artif. Ultramarine

Smalt

CoO & Al_2O_3 (Thenard's blue) much used in Coloring.

Note

$CoCl$ — has the property of being in neutral solutions & when dry — of a different Colors. If we dissolve $CoCl$ in H_2O — it gives us a red solution — with excess of HCl — it is turned blue. & when evaporated to dryness it is intense blue.

Ex

Lecture 101st

Co_2O_3. by gentle Heating of CoO NO₅ –
If pure heat the hydrated Oxide CoO HO with – NaO ClO₅ NaCl – we can obtain it – We know of no compds Co_2O_3, formed by it – it acts like MnO_2 when treated with acids (HCl &c) it gives off Clorine + forms a salt of the protoxide,
There are compds indirectly obtained having composition $Co_2Cl_3, 5.NH_3$ – the salt is insoluble + the Co can be obtained pure; formed in the acts in the separation of Co from Ni. =
There are a whole row of such substitution products. The one above quoted has a rose red Color.

945

Nickel does not form this comp'd – or I sh'd say this row of Compds. for many of them are formed + this gives us a splendid mode of obtaining Co free from nickel.

Sep. of nickel from Co

By glowing CoO in the air we obtain CoO, Co_2O_3;

$CoO Co_2O_3$

by glowing ZnO with $CoO NO_5$ – we can form a compd $ZnO Co_2O_3$: a green colored comp'd – a test for Zn.

If we ~~tr~~ treat CoO, with KCy we obtain a compd $3KCy, Co_2Cy_3$ → unstable.

note

important, for nickel only forms one compd – viz:– $NiCy, KCy$ – we are able from this behavior to effect the complete sep. of Ni + Co.

If we treat these two Cyanogen Compds, of Ni + Co, with HgO HO, the nickel salt is decomposed:– see KCy – NiCy / HgO HO } KCy, HgCy / NiO, HO

Co_2Cy_3, 3KCy does not suffer this decomposition but remains behind unaltered – the excess of HgO + the NiOHO is filtered + HgO driven off – the res. weighed.

Zinc.

Occurs in nature abundantly as Zn S – Zinc blende then Red Zinc Ore – ZnO, (not used for ...) Occurrence

Smithsonite (Spain) ZnO CO_2 + ZnOHO, ZnO SiO_2 Kiesel zinc – Willemite (ZnO 2 SiO_2) – are the best ores from which the metal is universally Extracted.

947
The mode of extraction is a very incomplete one - much being lost by its volatility - & its easy combustibility - & it is capable of many advantageous modifications. Coal reduces zinc - but zinc is such a inflammable metal that it cannot be produced in blast furnace like Fe. So it must be reduced in a closed retort & distilled over. in the distillation - it inflames - & much is thus lost - the figure illustrates the retort, & the distillation of the metal.

Note

Manufacture

Zinc so obtained is very impure As - Fe - & Pb. it is therefore redistilled, in order to purify it.

Zinc (pure) upon rapid cooling - can be obtain in crystalline leaves - ductile & malleable at about 100°. higher - 250°C - it is extremely brittle - Is the most expansible of metals - is very oxidizable by smelting it burns in air with a bluish flame forming. ZnO - resembles Mg - very much - Decomposes H_2O at high temps Ex or in presence of an acid.

Oxides

ZnO = Protoxide of Zn.
ZnO_2 = Super oxide "

ZnO.

by simple burning Zinc. Comes in this form in fabric. Called Zinc White. in Handel.

949.

Is poisonous – & very infusible – the use of zinc as household vessels is not to be recommended.

Note

HS no ppt in acid	NH_4S	NH_4OHO	NH_4OCO_2
	White ZnS	White – sol in Ex	White sol in Ex

Exc

NaOHO White Sol – in excess

$NaOCO_2$ – insol. in excess.

With HS, it is not precipitated from strong acid solutions (HCl or NO_5), but in neutral solutions or with weak acids A. 3.6. it is precipitated white.

$ZnOCO_2$ – by precipitating ZnO salts with $NaOCO_2$ looses its CO_2 by gentle Heat – it is generally basic ($ZnOCO_2$ + ZnOHO)

$ZnOSO_3$ – White vitriol in Händel. Crystallizes with 6HO isomorphous with FeO or $ZnOSO_3$ + 6HO quadratic.

Forms one branch of
the great group of iso-
morphous double salts
$NH_4O\ SO_3 + ZnO\ SO_3 + 6HO$.
&c, &c. **ZnCl**

ZnCl - by evaporating a
zinc salt in HCl - soluble
in H_2O - very deliquescent
volatile &c. Antiseptic - Use in
By adding a quantity to H_2O of HO baths
a HO Bath - the temperature
is elevated as the solution
becomes more concentrated - to 180°C.

$SbZn_3$ - (like Ammonia)
an alloy - of these two
substances - of no very
great importance - Cu + Zn Brass
Sb_2Zn_2 is also known -
but is only of theoretical
interest.

957

Indium

In — Discovered by the spectrum in SO_3 residues — comes with zinc ores — in almost infinitely small quantities as yet. Obtained as follows. We can obtain it from the manufactured zinc. Dissolve the zinc in HCl & digest with excess of zinc. Indium, Lead &c separate & are found at bottom as a black powder. The greatest part is Pb. It is dissolved in NO_5 &

Darstellung — treated with SO_3 — $PbOSO_3$ is separated — (as $PbOSO_3$) in form of an insoluble white Powder.

The solution is precipitated with NH_4O — ZnO

is dissolved up in excess – & $Fe_2O_3 \cdot 3HO$ with $In_2O_3 ? 3HO$ thrown down
Dissolve up in HCl & add SO_2 – FeO salt is formed – by adding BaO, CO_2 – the Indium is precipitated with the $BaOCO_2$ & the Iron is left behind. (Can be obtained metallic by KCy). Detected by its mono-Chromatic light – One intense blue line ~~light~~. Its oxide is volatile –
Metal much resemblance ~~to~~ Zinc – attacked easily by HCl. NO_5 &c,

InO – (or In_2O_3?)

White – volatile &c ~~(like zinc)~~,
few salts have been formed. & they are difficultly crystallizable.

953

Lecture 102nd.

Sn.

Sp. gr = 7.362. At. wght = 35.9

Tl.

Sp. grav = 11.862. At. wght = 204

Comes in nature in many Pyrites - Copper Pyrites.

Obtained from the product of the manufacture of these ores - Generally found in certain positions in the tubes (or ~~chimney~~ flues) of the SO_3 manufacture from these ores - Then it collects in form of a black powder - + from this by the process to be named - it can be manufactured by the pound, though but small quantities are in reality obtained. Dissolve in an acid &

SO_3 residues

We have only to place in the mother liquids - a large plate of zinc - this reduces first Cu, then Cd - & finally Thal. By this means we separate all Thallium in the form of a black metallic powder. By warming in SO_3 this metallic powder, only Tl, with traces of Cd goes into solution, by adding to this solution HCl, we precipitate the Tl in form of a Cloride - while Cd & Cu remain behind. It can easily be purified.

Manufacture

It is an easily reducible metal - in the stream - or by zinc - It is a singular metal - a great resemblance to Pb. & on the other

Properties

Op 5

pans, great resemblance to →
K. or Na – &c
Has a great affinity for O
+ in the air – rapidly covers itself with coat of Oxide,
but does not decompose
H_2O at ordinary temper.
+ only in presence of an acid.
in H_2O it always remains
glanzend, surface clear
+ metallic, for its oxide
is soluble in H_2O, + any
H_2O formed would be
soon dissolved off – leaving
the surface clear

Oxide

$Tl\,O$ = "Thallium Oxide
$Tl\,O_3$. = " Super Oxide

Both capable of forming
salts – well defined + crystallizable.

TlO.

Easily obtained by treating the Sulphate with BaOHO

TlO,HO. crystallizes with yellow color. (2 modifications exist). The Oxide is as before remarked, soluble in HO. is alkaline

TlO Salts Etc

TlO CO_2 - is soluble in HO - we only find such instances of solubility with the alkalies.

TlO SO_3 → dissolves in HO. under the [illegible] conditions mentioned before.

With $Al_2O_3, 3SO_3$ forms an Alum (TlO SO_3 + Al_2O_3. $3SO_3$ + 24HO) Forms SO, with FeO SO_3, MnO SO_3, an Alum &c it forms a whole row of isomorphous double salts, playing the part of Alkaline Sulphate.

957

Judging from the affinity of the Metal for O - + from the solubility of its Oxide - of its Carbonate - + Sulphate, + from the analogy afforded by its isomorphous salts (alums) &c we should place Tl. with the alkaline metals - (it forms too a salt TlCl, $PtCl_2$) → but the insolubility of its Cloride + Iodide + other behaviour - + its non decomposition of HO - we must place it ~~with the Pl.~~ [Tl] group

Tl.- better Classed with the Alkalies?

→

Tl Glass

Glass in which TlO plays the part of alkaline constituent - possesses great ~~refrac-ting~~ disper-sing power - + for certain optical purposes, could be found superior to all other kinds - the subject needs study - certainly deserves it.

Note

Tl I - is an exceedingly insoluble compound - best for quantitative deter-minations -

(Reactions

HS.	NH_4S.	NH_4O.	NH_4OCO_2	Ex
o	o	o	o	

Tl gives no precipitates here → the best reaction for Tl is the spectrum monocromatic light - ~~blue~~ green?

Tl is precipitated from its solutions by HCl - (like Pb) in much HO - soluble ↑ ← Ex

KI precipitates from solutions ↓ P of Tl salts - TlI a very insoluble compound - most insoluble of all.

With $PtCl_2$ in HCl solution a double salt $TlCl\,PtCl_2$ note completing the analogy with the metals of the alkalies.

959.

Poisonous Effects. The vapors of Tl are poisonous — + several cases are reported in which deaths has resulted from their constant inhalation.

TlO_3

Is likewise a capable base i.e. can form well-defined crystallizing salts.

TlO_3. If we add to a solution of a ~~Salad~~ salt of TlO — NaO, HO + lead Clorine through it — a salt of this TlO_3 is formed — it is tri-acidic.

Pb group

Comprises, Pb. Cu. Cd. + exe following Properties —

These metals either will not decompose HO under any circumstances, or only at a white Heat.

Pb. Lead

Known to the ancients, as its extraction from its Ores is accomplished most easily.

Occurs abundantly in Nature— As Galena— (Blei glanz). crystallizing in combinations of Regular System — ∞O∞ - O. ∞O, &c, then as Cerusit $PbOCO_2$ — as Pyromorphit. $3(3PbO\,PO_5) + PbCl$ — & as $PbOSO_3$ (isomorph. with $BaOSO_3$) — **Ores**

Manufactured as follows from PbS. (2 methods) One the precipitating process — **Glowning**

The Ores namely are mixed with Iron refuse — viz:—

$PbS + Fe = FeS + Pb$. by heating

2nd, Roasting PbS in air forms $PbOSO_3$ — coming into contact with PbS — this reaction ensues:—

$PbOSO_3 + PbS = SO_2 \uparrow SO_2 \uparrow + 2Pb$ — & this actually occurs.

961

This Pb is generally impure from Au + Ag.
By forming the acetate + recrystallizing repeatedly it can be obtained perfectly pure.

Properties — Pb has a bluish grey color — very soft — can be pressed + hammered — can be bent in all directions — can be obtained crystallized — (O)
Oxide is fusible — (those of the other group Fe &c are highly infusible)

note →

Pb in moist air forms PbO + PbO CO_2 — Roofs of houses an example —
Pb will not per se — decompose H_2O under any other circumstances — but in combination with Fe &c

when treated with an acid – a galvanic current is formed by which means HO is decomp. The use of leaden vessels in the household is highly dangerous – the metal being easily attacked, & highly poisonous in its salts – whole cities have been poisoned (Amsterdam)

Pb. poisoning.

Oxides

Pb_2O = Sub-Oxide of Pb.

PbO = ——→ " " "

PbO_2 = Super " " "

PbO.

Is formed by oxidizing Pb. in the air – or by glowing carefully the $PbO\,NO_5$ – forms a yellow colored powder – smelts easily & can readily be reduced by C – but not per se – Is very slightly soluble in HO – hence Pb

PbO

¶63

Poisoning -

(All the metals of this group are reduced to metals from their Compounds - by Metallic

Note → Zinc -)

HS.	NH_4S	NH_4OHO	NH_4OCO_2
White	white	white, insoluble in excess	white.)

Ex — NaOHO gives a precipitate Soluble in Excess.

SO_3HO gives even in dilute solutions precipitates of $PbOSO_3$

$KO2CrO_3$ gives a bright yellow precipitate of PbO,CrO_3,

Ex → Zinc reduces PbO Solutions a draft of Zn forms vegetation

Salts

$PbOCO_2$ - called white lead - most extensively + universally used in arts as a white Coloring matter — much superseded by $ZnOCO_2$.

Comes crystallized in nature as socalled Cerussit or White Lead Ore - Crystallizing Rhombic ($P\wedge P = 117^\circ - 118^\circ$.)

Occurrence.

Manufactured in the laboratory by Precipitating any soluble PbO Salt with a Carbonated Alkalie,

It is manufactured in many ways -

Manufacture

1st The process most generally adopted - Large plates of Metallic Pb. are surrounded with Moist Stable Excrement - many Plates of lead succeeded by alternating layers of this material - the Pb slowly oxidizes, & unites with the CO_2 liberated, by the decomposition of the Organic matter - after the lapse of

966

considerable time, the plates are removed + are found to be converted (for about 1/2 to 3/4 their thickness (according to the time) into $PbOCO_2$, this is removed + purified by washing — As it is generally sold — it is somewhat mixed with $BaOSO_3$, to increase its weight — while otherwise the expense of manufacture would be too great.

Lecture 103rd

Another method is to precipitate a basic Acetate with CO_2 Called called the French method - **Salts**

It is generally basic - Carbonate. It is only in traces soluble - by gentle heat it looses its CO_2

$PbO\,SO_3$ - is an insoluble powder - soluble in traces - by heating with PbS it is converted into $2SO_2$ & 2Pb - One mode of Pb manufacture.

$PbO\,NO_5$ - - crystallizes its NO_5 without HO - gives a pair of basic - Salts - they react slightly Alkaline. PbO is a base which has a great tendency to form such basic Compounds.

967

Crome yellow + Crome red are important salts for coloring. $PbO\,CrO_3$ by precip. by $KO\,2CrO_3$ or $PbO, 2CrO_3$ by treating this salt with concentrated NO_5 which seizes on some of base

Salts

→

$PbO\,SiO_2$ – used as a glazing for Earthen ware.

$PbCl$ – by precipitating a PbO salt by HCl in conc. solut. – We use Pb for many vessels in the laboratory – while it is very slightly attacked by acids + alkalies – for example in generating HFl –

⇛

PbI – precipitated by HI – a yellow solid – somewhat soluble in HO, cannot be used as a coloring matter on acct of its decomposability.

PbS. precip. by HS. a black solid – in Nature PbS Galena

Pb_2O -

If we heat Oxalate of Lead in a confined space - without presence of much air - this Compound of Pb with Oxygen is formed. A black Powder Pb_2O
By heating with an acid, a PbO, salt is formed & Pb is precipitated - It is as ~~not~~ a salt forming base - It is really a chemical compd. for, Hg. when mixed with it fails to draw out any Pb. which it would do from a mixture -

$Pb_3O_4 = (2PbO, PbO_2)$

Called Minium - Obtained by heating PbO - in presence of the air - by which it is partially Oxidized. Pb_3O_4

969

By treating this mennige with NO_5 PbO dissolves up - & the PbO_2 remains behind -

Ex

<u>PbO_2</u>

A superoxide - a good oxidizing - dissolves in KO HO & gives a crystallizing compd

By heating gradually it completely reduces to PbO,

Abtained by treating Mennige (Pb_3O_4) with NO_5 + filtering the insoluble PbO_2, It is a dark brown Powder,

Bismuth

Occurrence

Comes in nature - as native Bi - as BiS - (in Sachsen). The Ores are smelted - (the native one is here meant.) & the metallic Bi collects below in the hearth.

Is not very pure but contains Fe + traces of Cu. Purification by treating the finely divided metal with NO_3 - some (little Bi) is dissolved + all the impurities are dissolved out. So purified. Used in the composition of many alloys - among which is the important 'Fusible Metal'

Properties

A white metal with brilliant lustre - Can be obtained by slow cooling in large Rhombohedrons at Ordinary temperatures brittle Can easily be powdered, upon cooling it possesses the property like H_2O of increasing its volume | like H_2O

Is completely brittle but at 200° it is ductile - + thus wire formed at this temp - is for a time ductile - but becomes brittle

971

Crystallizes in Rhombohedrons (Isomorph. with As & Sb).

Does not dissolve in dilute acids - Does not decompose (Except at a white heat)

Note → HO under any circumstances.

Oxides

BiO = Prot-Oxide of Bi

Bi_2O_3 = Sesqui " " "

Bi_2O_5 = Per " " "

Bi_2O_3

Bi_2O_3 By dissolving Bi - in NO_5 we obtain a nitrate of this oxide - by glowing it remains behind - can readily be reduced by glowing in a stream of C & H. It is a yellow powder - comes in nature as "Bismuth Ocher"

Bi_2O_3-3HO - White Powder

If we add to a concentrated $Bi_2O_3, 3NO_5$ - Water - we get a white precipitate of basic Salt - Reactions Exp. Note

HS	NH_4S	NH_4OHO	NH_4OCO_2	$NaOHO$
copper brown	ditto	white insol in excess	Ditto	Ditto

Exp.

Just like PbO - in general -

If we add to such an excess of NaHO - & then add to it a salt of Bi_2O_3 - we get a black precipitate BiO. With PbO salts this reaction - Flame does not occur. - & it is a sharp qualitative test. Exp. Reactions

Bi is volatile - if we hold with asbestus Bi_2O_3 in flame (Reducing) + let the Bi_2O_3 deposit upon a cold porcelain dish, Then ⟹

973

by converting this into Iodide i.e. by burning an alcoholic solution of I upon it) we obtain a characteristic re-

Ex

action — By breathing upon it it easily dissolves.

With NH_3, it gives a yellow (red) color — an ammoniacal compound —

$Bi_2O_3\,3NO_5 + 10HO$ — Crystallizes

Salts

with HO — it is precipitated — a basic salt formed.

Bi_2Cl_3 — by treating Bi with Cl-gas — can be distilled at high temps. — forms compds with KCl, NH_4Cl &c — all crystallize

Forms an Oxychloride. Can be regarded as a chloride in which part of the Chlorine has been replaced by Oxygen.

If we smelt Bi_2Cl_3 + add to it Bi_2S_3 - we obtain a comp[ound] - $Bi_2\{^{Cl}_{S_2}$ - a cloride in which part of the latter substance has been replaced by sulphur

Bi_2O_5 -

If we lead into a suspended (mixture) of $Bi_2O_3 . 3HO$ in KOHO, Cl gas - we obtain this, compound - it is treated (cold) with NO_5 - + filtered + washed with NO_5 - HO + HO (We know of none of its salts) Bi_2O_5

BiO

Has a black color. Formed by precipitating it, from an Eo salt of the oxide Bi_2O_3 - by means of a mixture of SnCl + NaO HO

Very unstable, no salts known
has a tendency to oxidize itself.
We use the Bi metal mostly
for alloys - an important
alloy is Rose's fusible metal
which is composed of 1 part
of Sn - 1 of Pb - + 2 pts of Bi,
used extensively in soldering
It is characterized by the
property of easy fusibility; for it
fuses below the boiling point
of H_2O - (perform its fusion)

Copper - Cu

Comes in nature - (native).
at Lake Superior. - in Basaltic, Melaphyric rocks -
as Malachit, + Cupper Lasur
as Copper Pyrites - in Kup-
fer Sheifer - spread over
many square miles of country.

From the Oxides it is easily obtained by simple reduction with Coal - (ordinarily with a flux added) -

Manufacture

From the Sulphide the process of manufacture is complicated - Consists in Roasting the ore to oxidize them + then smelting them repeatedly with Siliceous Fluxes - By this means finally a tolerably pure Cu is obtained, but still containing more or less Sulphur + Fe

Lecture 104th

We must to obtain the metal from it's ores when it is in form of CuS, get the sulphides in form of Oxide — + the sulphides are for this purpose are repeatedly roasted in the air — this drives off much S, as SO_2 — + then when sufficiently so treated the ores are mixed with Coal + a fitting flux, + heated in the blast furnace, with the metal thus obtained, a process of 'puddling' is done ~~through~~ with which finally a tolerably pure ~~ore~~ metal is obtained, this is then thrown melted into H_2O, + allowed there to Cool.

By decomposing $CuOSO_3$ by the battery - or by reducing CuO with H, we obtain. metallic Cu The metal is of a deep red color - Crystallizes in O. (small) Is very ductile. Sp. grav. 8.94 Cu fuses very difficultly - & in fluid state absorbs O + N just as any other fluid - hence vessels cast of Cu - always have hollows upon their surfaces. ⟵ Is very stable in the air (when dry) - in moist air - Oxidizes itself slowly to CuO. NO_5 - is the best solvent. Dilute SO_3 + HCl do not attack it perceptibly. Conc. SO_3 dissolves it by evolution of SO_2. Cu Does not decompose HO under any circumstances.

979

Cu vessels — Vessels of Cu are dangerous if any acid substances are brought into contact with them — (So of fats &c)

Cu Oxides

Cu_2O = Sub Oxide

CuO = Oxide

CuO_2 = Super Oxide.

CuO

CuO — The chief oxide — formed by the oxidation of Cu — or by glowing $CuONO_5$ — or $CuOCO_2$ — or $CuOSO_3$ — it is a brown black powder — Easily dissolves in acid — is easily reducible by H or C. fusible at High temps. most of the salts are of a are of a blue Color — It is a salt forming Oxide — Reactions as follows.

HS	NH_4S	$NH_4O\,HO$	$NH_4O\,CO_2$
black	black	blue Sol. in Exc.	blue-green

With KCy, Fe_2Cy_3 it gives a red-brown precipitate very characteristic. All the metals of the groups above it reduce it from its compds. Iron placed in such a solution covers itself with a coating of Cu. **Cu** **Etc**

CuO HO - a blue solid - coloring matter

$CuO\,SO_3 + 5HO$ in tri-clinic system very stable - an important salt. Anhydrous $CuO\,SO_3$ is used as a dessicator of $C_4H_6O_2$ as it has the tendency to take up its crystals HO. **Note**

Salts

$CuO\,NO_5$ -- an important salt from which we obtain the CuO. Swinfurter's green - a beautiful green color

981

Swinefurter's Green

It is a mixture of the Acetate with the ~~acid~~ Arsenate of Copper, & much used in some countries as a coloring matter. It is exceedingly dangerous when not glazed.

$(\overset{''}{Cu}O, C_4H_3O_3 + CuO\,HO-)$ - Greenspahn used as a colorer.

Salts

Acetate (neutral) has a great tendency to form double salts - $(\overset{''}{Cu}O\,C_4H_3O_3 + CaO\,C_4H_3O_3)$

$CuCl + 2HO$ crystalline - soluble in HO - gives double salts with KCl, $NaCl$ &c. It is volatile - & is used on that account to test in the flame for the presence of Copper - staining it of a deep intense blue color - & giving a beautiful spectrum.

$\overset{I}{Cu_2}O$

Easily obtained by fusing 1 Equivalent of CuO with ditto of metallic Cu – again by adding to a solution of (a CuO Salt, T – + then) NaO HO, [not necessary to add Tartaric acid] + to this double salt – adding an organic Compound, best grape sugar, the CuO Salt is readily reduced & a sub-Oxide of Cu is formed precipitated as a red powder, can be filtered off & washed – insol. in Water but forms unstable salts. Ex

HS	NH_4O HO.	$NH_4O CO_2$
like CuO	white. sol in exces	yellow, sol. in exces.

NaO HO, yellow. Cu_2O, HO – NaO CO_2 – red-yellow, $Cu_2 O CO_2$ + CuO HO. Ex

Cu_2O Salts – partic – $Cu_2 Cl$ – are precipitated by HO.

983.

Ex With NO_5 the solutions of Cu_2O salts are oxidized to blue solutions of CuO, salt.)

CuO_2.

CuO_2 By treating suspended CuOHO in KOHO - + leading into it Cl gas - we get a red solution - $KO.CuO_2$? - from which however - the CuO_2 does not allow itself to be separated. (alloys)

Many important ^are formed from it (Cu) Brass - of Cu + Zinc, → German Silver → of Copper-Zinc, + Nickel - Alloys then Kammonew Metal - of Cu + Sn

→ Bell Metal " Cu + 2Sn

→ Mirror Metal " ditto

→ Medallion Bronze " "

→ Bronze. " Cu, Zn, Sn,

→ Brittannia Metal Cu Sn, Sb.

Cadmium

Comes associated always with Zinc Ores - seldom independently - When first reduced it being more volatile than Zn it settles first upon the Oven walls in form of a brown dust, which when collected is found to be rich in Cadmium. This Cd containing some Zinc 30%. is next mixed with C, + heated in closed ~~oven~~ retorts + gently heated - the Cd - distills over first - being much more volatile than Zn - by repeating this distillation several times - always - rejecting the last parts - it can be obtained tolerably pure - So comes in Handel.

Occurrence

Free - ing - form Zinc.

985

this contains yet several %
of zinc - from which it
Chem. can easily be obtained perfect
Purif- ly free -> i.e. by dissolving
ication in NO_5 - glowing - + reducing the metal oxide in
a glowing tube by H, gas
the H - will not reduce the
zinc - + the Cd metal distills
over perfectly freed from
it - **Properties.**

Resembles Sn in appearance,
but in prop. more Zinc.
It is much harder than Zn -
Ex + still more volatile - i.e.
flame more easily to distilled. Stable
test Heated it burns in air
~~forming a brown Oxide~~
It is the only metal of this
note group - which will decompose HO in presence of an acid

Cd + Oxygen

CdO = Oxide —

Cd_2O = Sub Oxide (?)

CdO.

A black powder – crystallizes in O – is a salt forming base – Is non-volatile – Its Hydrate is White – & easily soluble in acids. Most of the Cadmium Salts of CdO – are colorless. Discussion next lecture.

Lecture 105th

CdO.

Easily formed by glowing a $CdONO_5$ - of non-volatile forms many salts -

CdO / Er

HS	NH_4O HO	NH_4S	NH_4O CO_2
yellow	white sol. in exc	yellow	white insoluble.

Salts -

$CdO\,HO$ Easily looses its HO on heating - So with $CdO\,CO_2$

$CdO\,SO_3$ - formed as usual - singular isomorphous with $3\,YtO.SO_3+8HO$ (Dr. La Er.)

Salts

$3\,CdO.SO_3+8HO$. for a time officinal.

$CdCl$ - is a Cloro-acid - & unites with basic clorides -

Double Salts

$BaCl, CdCl+4HO$.
$2NaCl, CdCl+4HO$
$SrCl, 2CdCl+7HO$
} double salts

& other clorides -

CdS - much used in oil painting, a stable color. Ex

Cd_2O

Formed by treating (glowing) the oxide with excess of Cd.
$CdO + Cd = Cd_2O$. Ct ←

forms no salts - by treating with acids a CdO salt is ~~action~~ formed + Cd is separated - That is not a mixture nota is proven by the fact that mixture Hg will not extract from ~~cadmio~~ it metallic Cd.

The next two metals, are characterised (as well as that the gold group), by simple heating - they have the property of giving up their O + of being reduced to metals. They will under no circumstances decompose HCl.

Hg.

A most invaluable metal to the Chemist + Physicist generally found in nature as Cinnabar HgS - then mixed with HgS as Hg.

Hg

We have only to mix the ores with CaO, + glow - + obtain the metal

Gewinning

$$\left.\begin{matrix} HgS \\ CaO \end{matrix}\right\} \begin{matrix} CaS \\ Hg + O \end{matrix}$$

It is distilled in closed retorts - it is hence somewhat impurified. But, we have only to treat the metal with Dilute NO_3 - allow it to stand - shaking up frequently in contact for some days - + the metallic impurities are completely dissolved out. So also some Hg.

Purification with $HONO_5$

Properties.

Pure Hg can be easily detected by a drop of it leaving no track - + rolling freely + clean upon a porcellain plate — Hg

The knowledge of the properties + laws of gases - were only detected by this ~~gas~~ impt. metal

Sp. gravity at 0° = 13.596

Melting Point = -39.44°

Color. bet. Sn + Steel - Seguedat Ord. C.

Does not decompose HO under any circumstances

With SO_3 it acts like Cu; viz:—

$Hg + 2HO\,SO_3 = HgO\,SO_3 + H_2O_2 + SO_2$

Vaporizes at 360° - the vapors are very dangerous.

By constant shaking - Hg is converted into an infinite number of finely divided particles - Medicine - with fat - Hg Ointment

Oxides

Hg_2O = Sub Oxide.

HgO = Prot Oxide.

both salt forming bases

Hg_2O + HgO Manuf. — If we allow NO_5 in the cold to act on Hg (in excess) – a salt of Suboxide is formed $Hg_2O\,NO_5$ – – $Hg_2O\,HO$ precip. out by $NaO\,HO$, If the conditions are reversed – (i.e. excess of NO_5 + heat – $HgO\,NO_5$ – is formed, $HgO\,HO$ → precip. in same way.

HgO

By glowing weakly $HgONO_5$ – so that all NO_5 is removed.

HgO + Salts — By precipitating $HgCl$ by $KO\,HO$, + washing well – the HgO first obtained is red – that obtained by precipitation with $KOHO$ is yellow.

There are ~~yellow~~ therefore two modifications of HgO

Oxalic acid ~~is~~ instantly attackes the red modifi-cation, while it only dif-ficultly + slowly attacks the ordinary red oxide-ic. — Differences bet. them

HgO, Reactions—

Best obtain the salts—by or from the $HgONO_5$—

HS	NH_4OHO	NH_4OCO_2	KOHO	Reac-tions
black	white Sub. prods	ditto Sub. prod	yellow HgOHO	

When precip. by HS—the sulp-hide has the property of for-ming double salts (white compds) z. b. (HgS, $HgONO_5$)—decomposes with Excess. into HgS. (Note with HS)

There is one method of detec-tion for Hg salts—that is infallible—We mean the ~~process~~ by which Hg is obtained. As follows—: Best test

993

Ex. bis – by mixing the suspected salt with $NaOCO_2$ + heating – the salt is reduced – metallic Hg is sublimed in the upper part of the tube + can be examined.

HgO, NO_5 — is a syrup thick fluid – (exists also tri-basic) by Heating decomposes into $Hg + O + NO_2 + O_3$ + etc.

$HgO\,SO_3$ – is a white solid – by mixing with HO – it is converted into free SO_3 + a yellow basic salt – which gradually decomposes,

HgCl – Corrosive Sublimate by leading Cl over Hg – or, better – by dissolving in Aqua regia – formed now as follows – by distilling from a mixture of Hg, NaCl + SO_3 (some

$MnCl_2$ to prevent any Hg_2Cl forming).
Much used in medicine
is a Cloro-acid - forms double
salts - Can easily be reduced
to Hg_2Cl.
HgI, by precipitating Hg (II)
salts by KI solution a — <u>HgI</u>
beautiful red color - by
subliming it is converted
into a yellow modification
forms double salts - (HgI
KI) - Dissolves in KI.
Used to test for Sodium - <u>Ecc</u>
the bright red salt appearing in Soda light - <u>yellow</u>.
HgS - upon precipitating | <u>Ver-
million</u>
black - If, however, we digest
for several hours with
KS, it turns all of a
sudden to a bright
red color characteristic of it.

995

HgS — Called 'vermillion' — is difficultly soluble — Used as a coloring matter —

Uses — Used in coloring 'Sealing Wax', &c.

HgCy — HgCy — a poisonous salt — forms double salts with basic Cyanides with KCy &c

Hg_2O

Hg_2O — By dissolving Hg at low temperatures — & in ~~excess~~ presence of Excess of Hg — By precipitating this Hg_2O, NO_5 — with KOHO we obtain it in form of a black powder — decomposible in air into —

Hg + HgO.

The reactions are entirely different from those of HgO

HS	$NH_4O\,HO$	$NH_4O\,CO_2$	$NaO\,HO$ (Me?)
black	greyblack	ditto	ditto

Easily distinguished from HgO salts by the black color of its precipitates, + from Hg_2Cl the fact that – HCl will precip- precipitate Hg_2Cl from solutions of this Oxide – NH_4O turns it black (Hg_2O) by HCl Exc

With SnCl – it is reduced from its salts – to metallic Hg. Exc

$3Hg_2O, NO_5, 3HO$
$4Hg_2O, 3NO_5 + aq$
$5Hg_2O, 3NO_5 + 2HO$
$Hg_2O\ NO_5 + 2HO$

Salts of $Hg_2O + NO_5$

Calomel Hg_2Cl – analogous to Hg_2O – an ~~important~~ salt in medicine – formed By treating HgCl – with metallic Mercury – as follows →

997

Manufacture	$2Hg$ $2SO_3$ $NaCl$ aq	Hg_2Cl $NaOSO_3$ SO_2 + aq

or. from HgCl - by intimately with the necessary amt of Hg - + subliming the mixture.

Ex → With NH_4OHO, it becomes black - (Hg_2O being separated)

Alloys so called Amalgams — Hg forms many alloys which we call - to distinguish them - Amalgams; so we can dissolve K, Na - Sr. Ca Ba &c &c in Hg - + some crystallize out of the Hg solution in beautiful crystals - there being an actual chemical compound formed.

It unites with most metals in all proportions - the state liq. or solid depends from Hg.

Lecture 106th

998

Analysis of a Russian Plat. Ore.

Pt. Ore

Plat	= 84.30
Pd	= 1.06
Ir	= 1.46
Rh	= 3.46
Ru	= Trace
Os	= 1.03
Fe	= 5.31
Cu	= 0.74

Gangue

FeO, Cr_2O_3	
FeO Fe_2O_3	
FeS_2	= 2.64
CuS FeS	
Osmiridium	
Pyroxenic Ingred	
	100.00

999

Ag

Comes Native in nature oftener as Rothgültig Erz. $3Ag_2S, SbS_3$(?). a fine red ore — then as Tetradymite - Fahl-erz a Sulph-arsenide - complica. As Melan-Erz, as Amalgam, (" AgCl - AgBr). Often associated with Cu + Pb.

Ag.

Those Lead ores (oxides) containing Ag, are treated when in a state of the finest powder - with NaCl + heated thereby - + the decomposition which takes place is the following - the principle point being the formation of insoluble AgCl. This is now mixed with Fe, in a finely divided state? + the foll. takes place

When much Ag is present

$Fe + AgCl = Ag + FeCl.$

a constant revolution being kept up, until all Ag is reduced, + Fe is present as FeCl. Then Hg is added in excess it dissolves out all the Ag + it is then distilled over in closed retorts, the Ag remaining behind. (first pressed through a fine porous leather!) This is the process pursued when little Pb or Cu are present + much silver. This is called the 'Amalgamating Process' — When little Ag + much Pb or Cu (impure) is present — the ore, is after roasting, Reduced — the Ag (+ Au) going into the Metal — The Pb — is then placed, in appropriately constructed hearths, + in a stream of air — smelted

{Then Amalgamates}

(a thick alloy remains behind)

1001

the Pb readily oxidizes itself + the PbO being fusible flows off, + at last the Ag (+ Au) is left behind in form of a metallic globule — free from Pb — (+ Cu) — This is called the 'driving off' of the Pb — This process is always adopted when much Pb. + little Ag (+ Au) is present; + is resorted to — whenever the lead ores are rich enough in Ag, to repay the trouble of fusion.

When much Pb ore is present

To purify the Ag completely we have only to dissolve it in NO_5 + to precipitate it — with HCl — AgCl is then washed out by decantation + reduced with $NaOCO_2$

Purification

$$\left.\begin{array}{l} AgCl \\ NaOCO_2 \end{array}\right\} \begin{array}{l} NaCl \\ Ag + O + CO_2 \end{array}$$

Is a pure white metal - of splendid metallic lustre, is exceedingly ~~stable~~, is very ductile Properties + harder than Au. (400 feet - from 1 grain)

Possesses like Cu - a great absorptive coefficient for certain gases, for O particularly therefore - when we reduce Ag before the blow-pipe, it absorbs O, like H_2O, + upon cooling - the gas escapes - giving the metal a peculiar porous surface. Note

It - is somewhat of a yellowish when deprived of its lustre - Is very ductile, can be drawn into the finest wire.

Will not decompose H_2O under any circumstances

NO_5 - is the best solvent -

With SO_3 it acts like all the metals of this group.

1003.

Oxides

Ag_2O = Oxidul - Sub Oxide
AgO = Oxyd - Oxide
AgO_2 = Super-Oxide.

AgO

AgO | Easily formed by dissolving Ag in NO_5 + precipitating this Solution by KOHO.
Easily reduced by gentle heat or by Organic bodies - to Ag + O - Is a salt-
Salts | forming base.

Salts.

$AgO\,NO_5$ - Can be smelted without decomposition. It is much used in Surgery as a Corrosive.
Is easily soluble in HO. - but is decomposible in light By highly heating it is Extensive decomposed.

AgCl - by precipitation of AgONO₃ - by means of HCl, HO, Is a white solid - insoluble, reduced by Organic bodies. Changes its color in the light, passing into purple - It is much used in Photography - exposed on a thin film, it rapidly changes - to black Ag_2Cl being form'd. $NaOS_2O_2$ (insol - in $NaOS_2O_2$) the unchanged AgCl - is dissolved off by Hypo Sulphite

AgCl

dissolves it

AgBr &c similar - but somewhat yellow in color,

AgI - similar to AgCl - all these salts (Halloids) are soluble in NaO, S_2O_2 - note

Silver plating consists in coating Cu, Brass &c with a coating of Ag - it is accomplished either by galvanic deposition or by rubbing an Amalgam & heating.

1005

Salts Ag S. by heating a mixture of Ag + S. or better by precipitation of a salt of AgO by HS.

note Is a conductor of Electricity + is ductile, + has some metal lustre

Reactions.

Ex

HS	NH_4OHO	KOHO	NH_4OCO_2
black	white Sol in	brown	white. (Sol in Ex)

HCl is the best reagent for Ag salts – the slightest trace of HCl – will give us a precipitate with a silver salt – insoluble in NO_5 but soluble in NH_4O + NaCl – its property of turning brown in light. is characteris. for Ag.

Ag_2O.

Ex If we take Rochelle salt – which contains SbO_3 – + add Excess of KOHO – + then a AgO

Ag_2O salt – SbO_5 is formed at the

expense of the Oxygen of the AgO - Ag_2O is precipitated black - very insoluble + unstable, Ag_2Cl is formed upon exposing a surface of AgCl - to the action of light - hence the rapid turning black of a film of AgCl on exposure to light - the art of Photography depends upon this property - Photography

For the quantitative determ. of Ag rapidly, a mode of anal. called Assaying is used. Peculiar weights + values are used in different countries. Approximately - Alloys of different values - in convenient form - are used - + by the color of their "streak" compared with the tested alloy - the Assaying

Ex.

Ex.

1007

purity of the test is judged of.
Cupellation is old - & the best
Cupella- method - a "Cupel" is made
tion of "bone ash" (Sheep's), & on this
a small (weighed) quantity of the, to be
tested silver is placed with
3 times wght of pure Pb. & heated
to white heat in the muffle -
(with air stream) PbO - with
Titrating CuO is absorbed, & the pure
Ag, com is left behind &
weighed.
In Titrating the test (a wghd quant) is dissolved in NO_3 -
& titrated with a normal
NaCl Solution - or vice
Ex versa, the test dissolved
in 1000 C.C. (or 500) & a
normal wght of Pure NaCl
titrated with it - both cases
will do as well.

Platin group.

A great amount of confusion exists with regard to many metals of this group - partic. Note the modes of their separation atomic weights + Sp. grav's.

Character is - this - that the Oxides - + their Salts - are of very little importance - while the Clorides, are the most important.

The metals possess the common property - of not decomposing HO under any circums.

Au comes pure in nature free from the other metals. The other metals however - (Pt Ru, Rh, Os, Iro) always come mixed with one another.

It Crystallizes in Regular System - in O + ∞O - planest.

1009.

Occurrence

Occurs mostly native, in veins in quartz. In California - Australia, Urals &c, loose in sand &c. Traces are gained by amalgamation.

Separation from much Ag

Separated from Ag by simply dissolving up in conc. SO_3 - the Ag is left behind undissolved.

Much Au with little gold is dissolved in Aqua Regia is treated with evap. + heating.

Sep. with Hg.

In the gold sands the sand, from which all large particles are sep. is treated with Hg. + thoroughly mixed. Then as with Ag - the Hg is distilled off + the pure Au left behind in the retort - freed from Ag see above.

$AuCl_3 + 3C_2O_3 + 3HO = \{ 6 ... + AuCl_3 =$
$Au + 3 ... + 6 ... _2 \}$ [illegible]

1010

To purify it - we do best to dissolve in aquaregia - evap Reduce orate + treat with Oe - with alic acid - $AgO + C_2O_3 =$ $\bar{O}$ - $Ag + C_2O_4$

Has a beautiful rich yellow color - has no absorptive coefficient - is exceedingly ductile - can be hammered into leaves so thin that it is transparent, light is green through it, is perfectly stable in the air, pure gold will keep bright for years. **Properties**

Is not attacked by HCl - by NO_5 - - by H Fl - only in a fluid containing free chlorine - is it attacked - **Note**

There is formed of course the Ter cloride of gold $Ag_2 Cl_3$

1011.

Lecture 107th

	Atom. Wght	Sp. grav
Pt	98.56	21.3
Pd	53.23	11.8
Ir.	98.56	17.62
Rh	52.16	11.00
Os	99.40	21.4
Ru	52.16	8.6

Au will not easily be acidized - (Au + S smelted together will not be attacked With plate or leaf metal, the

Ex fact as to whether it is Au or not, can easily be tried, by heating it with NO_5, or with HCl - imitation gold leaf - of Cu &c - are instantly dissolved - whereas a leaf of the genuine, metal will not be in the least affected.

Sp. grav = 19.8 ←
By dissolving in Aqua Regia we obtain $AuCl_3$ – by evaporating to dryness + heating (but not to glowing) gives us AuCl – by still more heating metallic Au, is left behind.

Properties

$AuCl_3$ forms many beautiful double salts – with Alkaline Clorides. (AuCl is insoluble in HCl)
These Clorides – correspond to two Oxides – but neither of them are of any importance – as no crystallizable salts

1013
with KOHO.
It forms no salts.

$Au O_3$

Obtained similarly by treating $AuCl_3$ solution with an alkali or alkaline carbo-

$AuCl_3$ nate ($NaOCO_2$) + boiling. Is very weak in its affinities.

Au in its salts, is easily reduced to metallic state by evaporating them with Oxalic acid.

$FeO SO_3$ — precipitates me-

Ex tallic Au from solu-tions — in form of a dark brown powder.

$SnCl_2$ + SnCl — gives in solutions of gold a deep

Ex purple precipitate — called Purple of Cassius — much used in glazing, composition

doubtful - (SnO - AuO?) ver-
Similar to silvering - but golding methods are more numerous - 1 - By Reduction - (Steel) from solutions by simply plunging it into $AuCl_3$ solution - again by galvanism

With NH_2OHO - au salts are precipitated - it appears to be a substitution of H6 Au by Au - & is easily exploded with great force - by friction or heating. Explosive

H_2S, gives - a black precipitate of AuS_3(?) - which is soluble in KS or NH_4S - hence a sulpho-acid.

Platinum Group.

These metals occur together & the greatest difficulty attends their separation - for, Difficulty when impure any of the metals give entirely different reactions from those which they give when pure.

1015
Their separation is therefore attended with amazing difficulties – & requires – Jobian patience –
⟶ They are found in Brazil – in the Ural. California – Plat – is found there with gold & in the sands with gold.
(Fine analysis, see 998)

Separation

Note We can just separate them. fine the Platinum heat with Aqua Regia; the Pt chiefly dissolves out, but, likewise, the other metals in traces. (the residue
⟶ contains little Platinum & much Os & Ir.
Solution = a. Residue = b.
Solut. contains the other metals also – but in traces.

Reduce the solution a, with Fe - Pt - Pd + Rh is contained chiefly in the Residue (C) These metals all give Clorine Compounds which have very analogous properties.

i.e. - KCl; $PtCl_2$ } bicloride
KCl, $IrCl_2$ } double
KCl, $PdCl_2$ } salts

+ again $3KCl$, Ir_2Cl_3 sesqui ditto

These compounds all the Pt metals enter into, except Rh - which does not form this bi cloride double salt These compds (bi-Clorides) possess a property by means of which we are enabled at once to separate all the Platinum chemically pure. i.e, Form the double Cl, compds as usual from the reduced metals, Digest with KOHO-(in pres. of Excess of KCl) the foll. ensues →

Pt Separated

$KCl, IrCl_2$	KCl
$KCl, IrCl_2$	KCl } Ir_2Cl_3
$KCl\ IrCl_2$	KCl
$KCl, IrCl_2$	KCl
	KCl } Ir_2Cl_3
	KCl
$2KO\ HO$	$KOClO$
	KCl

By treating with $C_4H_6Cl_2$ we decompose the KOClO into KCl, The Sesqui Cl. are characterized by giving no precip. with KCl - while the bi-Clorides without Exception do.

We therefore, treat the solution with KOHO to form the Sesqui Clorides, + thus succeed in decomposing all the bichlorides Except that of $KCl, PtCl_2$

1018.

Platinum does not form a sesqui-cloride – & by then adding KCl to the solution Only the Platinum is precipitated – chemically pure the Pt is now completely separated. Is a bluish-white metal – Sp-grav = 21.3 – a equiv. Wght = 98.56, is very ductile Platinum Smelts only in the fiercest Properties. heat, we are able to produce. Can easily be hammered – When impure – (Ir) it is much more brittle. Like Au – it is not attacked ~~not~~ easily – by Phosphorous – (some metals) – Only by fluids giving off Cl. Hence its adaptation to the manufacture of Capsules being almost infusible – Russia manufactured Coins from Pt – (has now ceased)

Coins

1019

$PtCl_2$

Formed by dissolving Pt in aqua regia. It has a great tendency to form double salts, with alkaline Clorides → By heating forms first Pt Cl + ^by still more metallic Pt.

$PtCl$ –

Formed as above mentioned; a green-yellow Compd. Insoluble in HO. We know of no separation into Sesqui Cloride; upon which depends the separation of Plat. from the other Platinum metals.

Er Cy + Pt Cy = } Isomorph salts
Yt Cy + Pt Cy = } Trichroismus
beautifully crystallizing salts.

1020

<u>Eisen-Rückstand</u> (Free of Pt)

In this then is contained no Platinum, but contains much Plad. + Rhodium.

The affair is smelted (fused) together with Zn + ZnCl (ZnCl used for the purpose of dissolving up is basic ZnCl any Oxide which might form.

If we dissolve up the Zinc all the Plat-metals are left behind, there are also present Cu + Pb &c (as impurities of zinc). HCl will in presence of the Pt metals dissolve up the Cu + Pb — Dissolve owing to the presence of in HCl Electric Currents by Contact.

Dissolve the rückstand in Aqua-Regia which is readily accomplished.

1021

~~(Pt)~~ Pd, Ir - + Rh are contained in Solution

Ru - is in the Rückstand

First separate the Pt - if any be present. by the mode described before,

Lead Clorine into ~~the~~ Solution, Pd + Ir are precipitated as Biclorides

$KCl, Pd\,Cl_2 \rightarrow$ (dissolve
$KCl\,Ir\,Cl_2 \rightarrow$ up in)

Rh does not form a bi-cloride - hence it is left behind as a solution of Rh_2Cl_3 - chemically pure.

By precipitating the solution with K I. <u>in the Cold</u>, we precipitate all the Pd as Pd I. Care must be taken to avoid excess & is accomplished by taking with a capillary tube a part of the fluid when near the right point — & testing it with KI. We then have the Ir alone by evaporating & adding Aqua Regia — we can Crystallize the Salt KCy Ir Cl_2 out —

No Excess

Rd.

Atomic Wght = 53.23 Sp. grav. = 11.8.
Dissolves in NO_5 — ←
Is attacked when highly Pd heated & forms RdO, (looses it again on cooling?)

Rd Cl_2, be cloride
Rd Cl — cloride

The double salts of RdCl_2 with Alkaline Clorides — we obtain double salts as with Pt but they are

1023

readily distinguishable from the corresponding salts of Pt. by their fine red color—(those of Pt being of a canary yellow color)

Lecture 108th

Palladium has one property, in which it stands alone i.e., Alloy. It can form with Hydro- of Pd + Hem- an alloy in all pro- portions- Proving that H- is a metal - but in a gaseous form. ⟵ H.

We obtain Ir + Rh best from the Rückstand C,

They are both insoluble, + they are ~~mixt~~ mixed with a 'Basic Cloride' BaCl, (or NaCl) &c, + just under the glowing heat - Cl gas is lead over it - there is form- ed - the well known double Clorides - which are read- ily soluble in Water (the excess of Cl - is disolved by absorption in KOHO). Note

1025

This is the only way of getting these metals in solution, for no acids - not even Aqua Regia will affect them. By adding $HOSO_3$ we can separate completely the solution from BaCl - & by pains taking the proper amount of SO_3 can readily be added to just precipitate the BaO, - from this solution - the Rt is separated. as before given - by digestion with an alkaline Hydrate & precip. by KCl.

Diplomo again. The residue containing Ir + Rh is reduced with H gas. again treated with BaCl - & the Ir separated from Rh. as follows:-

1026

The fluid is treated with HCl - concentrated, & then in the cold - an excess of (acid) NaO SO_2 - added - & thus allowed to stand for several days the Rhodium gradually separates in form of an amorph. yellow double salt of Soda & Rhod

Sep. of Ir + Rh

$NaOSO_3$
$Rh_2O_3SO_3$?

Properties.

Rh.

Is a grey white metal - which is ductile, malleable - & capable of great polishing
Is completely insoluble in acid to dissolve it, we must mix it with NaCl & lead a stream of Cl gas into it, (as mentioned in the Separation of Rh & Ir.)
Rh. however - in spite of its insolubility & invulnerability to acids is easily oxidized in the air

1027.

Upon heating moderately the only Union of 1 Cloride & 1 Oxide - of any importance although others exist. The behaviour of many compounds - is yet - doubtful - as the greatest discrepancy exists between the statements of some Chemists - Bunsen's method of separation the only one.

Ir.

Ir Easily obtained pure - Contained mainly in Residue - Fused with BaCl - & upon solution Reduced with H - & separated from Rh - as above mentioned -

Difficulty Reduced It is only reduced with difficulty & slowly by H.

Ir possesses the property of being converted into Sesqui Oxide - when a solution of Bi Cloride - is treated with an alkali. The reaction being noticeable by an instantaneous decolorization of the solution. Ex-

Properties

It is the most unmanageable metal of which we know, Cl don't attack it, we can set Cl free upon an Iridium plate, in the battery, & it will not be in the least affected, very difficult to alter after it has once been placed in metallic state

Oxides

IrO, Ir_2O_3. IrO_2 - IrO_3 - (& Corresponding Clorides?)

1029.

Ru.

Remains behind in the Residues after having separated all, the others, by extraction with Aqua Regia & $BaCl$.

Note → By smelting the mass with $KO\,NO_5$ & $KOHO$ — we obtain a peculiar Oxide of Ruthenium, which forms a soluble compound. & can be thus be obtained pure. Much like Ir, in its properties. Is exceedingly hard; finds an extensive use in the arts.

Use — particularly in the tipping of gold pens, for which its hardness peculiarly adapts this body.

(Ird. Containing some Ru)

1030

Os -

Possesses peculiar properties — Is very volatile, forms, by treating the original ores with Aqua Regia - Osmic Acid - which volatilizes with the fumes of the acid — if these are passed into a vessel of NH_4OHO - it will be completely absorbed - forming a compd → ($NH_4O\ OsO_4$?) From this - H - will readily reduce it.

Os

Combine it with, NH_4OHO

Os

Is a black metallic lustrous substance -, it volatilizes at a high temp. before smelting. Burns when at red heat, to OsO_4 in the air or Oxygen - OsO_4 has a fearfully irritating smell

Ex

Smell of Os

1831.

resembling very much that of Clorine → but even more dangerous in its Effect.

Osmic Acid = OsO_4 - formed upon heating an Osmium compound → with Aqua Regia - is exceedingly volatile & can be volatilized from spot to spot, like Hg - or As. Is easily reduced - by all Reducing agents (Fe_2, & C_2, etc. Has the above mentioned terrible smell, which is utterly unendurable to the sense & membranes.

OsO_4

→ Os forms the following Salt forming Oxides OsO, Os_2O_3, OsO_2, OsO_3, OsO_4 Many salts have not been formed - the last compds forming crystallyzing compds

The plat. metals, have one property in common, which no other group posess - viz:- That they are reducible to metals, from their salt solutions by H gas → by such treatment we can readily separate them from other metal groups. A singular regularity is observable in the Sp. gravs + Atomic Weights of these bodies - the same standing 2 + 2 in relation as 1 to 2.

General property of the Pt. metals

At. Wgts & Sp. grs

Mo + Wo. group

Occur sparely in nature → + are reckoned to the rare substances, the group is characterised by the absence of positive metallic character - (as likewise the group following.)

1033

Lecture 109th

Wo + Mo.

Elements.	Sp. gr.	Atom. wght
→ Wo	17.5	92
→ Mo	8.6	46
→ Ti	?	25
→ Ta	10.	91
→ Nb	6.5	47
→ Va	?	51/371

Wo.

Occurs as MnO (FeO) WoO_3
The acid can easily be obtained, by treating with an acid - (aqua Regia) - the acid remains behind - The metal can be reduced from it by H. or C.
It is a grey metal - very brittle, + has a high sp. grav. Burns in air - readily upon heating - to WoO_3

Oxides.

WoO_2 = Bin Oxide - unimportant

WoO_3 - Wolframic Acid! (Tungstic)

WoO_3.

Is a yellow - fine powder
Not alterable by heating -
gives with KO HO +c+c - fine WoO_3
crystallizing salts, of singular composition - There are
three rows of Salts :- viz :-

$3KO, 7WoO_3$ = 1st. Isomorph. row.
$KO, 4WoO_3$ = 2nd " "
KO, WoO_3 = 3rd " "

The middle acid is very
different in properties to the
first + last - we call it
Wo_4O_{12}.

If we add an acid ~~to~~ any
Salt of WoO_3 - we obtain a Ex
precipitate of white WoO_3, HO
white, by heating looses HO + becomes

1035

Ex yellow.

Common Property of this & the next group (i.e. Va + Sn) — All these metals possess the property - like Silicium of entering into most intricate & unusual compounds - & all stand on transition ground - between metals & Metalloids.

If we heat moderately the salt - 3NaO, 7WoO$_3$ - it falls apart into 2 salts - one soluble & the other insoluble, viz:-

(Not precip. by acids - & Sol. in HO)

NaO, 4WoO$_3$ } 3NaO, 7WoO$_3$
2NaO, 3WoO$_3$ }

Meta-Wolframic Acid — The acid 4WoO$_3$ (or Wo$_4$O$_{12}$) possesses very extraordinary properties - different from WoO$_3$ - so that we must look upon it as a polymeric modification of WoO$_3$ with the composition Wo$_x$O$_m$.

Ex

Converted into ordinary WoO_3 by evaporation almost to dryness with conc. SO_3 -   Ex

It (ordinary WoO_3) has the property in common, which SiO_2 possesses - of forming salts of most uncommon constitution -

WoO_2

By glowing WoO_3 moderately in H gas or with C. Is utterly indifferent - insoluble in acids + bases. It unites however with WoO_3 forming a blue colored fluid - Precipitated black - by NH_4S upon addition of an acid.   Reaction Ex

MoO - Occurrence

Occurs as MoS - (appears like Galena)

" " PbO, MoO_3 Yellow Pb Ore.

We make use of MoO_3 salts -

1037.

in testing for PO_5 – ($NH_4O\,MoO_3$)

Like WoO_3 the acid is separated – by digestion of $PbO\,MoO_3$ with a strong acid to evaporation, nearly to dryness, on the H_2O bath – HCl is best.)

Separation of the MoO_3

MoO_3 separates + $PbCl$ formed(?)

Or, from MoS_2(?) by volatilizing the same – in a strong heat + in presence of air – SO_2 is formed, + MoO_3 – which condenses in large crystals in the top of the Iron Tube

Can be reduced by H. gas (or by C). NO_5 – attacks it.

Oxides

MoO, + MoO_2 = Unimportant.

MoO_3 = Molybdic Acid.

Molybdän Same

Obtained from the Ores as above mentioned – by treatment with a strong acid.

It is a white powder - insol- 1038.
uble in HO - but soluble in
NO_5 - (+ HCl?) - volatilizing at a
red Heat -
Like MoO_3 - acids, give a white Reaction
Precipitate - but it is easily soluble in excess. This acid so- Ex
lution is easily Reduced to
MoO_2, MoO_3, by Sn. (or SnCl) -
blue color - but the reduction
goes on farther & MoO_2 + finally
MoO. is formed - brown in color
MoO_3 precipitated brown by Ex
$KCy\ Fe_2Cy_3$. The Characteristic Reaction is - that $2NaO\ HO$
PO_5 - - will give a yellow precipitate MoO_3 yellow - by formation of $3MoO_2, PO_5$ ← Ex

MoO_2

Formed like WO_2 is formed.
Dissolves in acids but
gives no Crystallizing Salts -

1039

Combines with MoO_3, to form the compd. (charac. of all this group except Sn + Ca) MnO_2, MnO_3.

note →

MoS_2 (streak) MoS_2 occurs in Nature as Molybdäne glanz, — looks much like PbS — but the streak is green.

Sn Group.

Sn is the only metal universally distributed. The characteristics approach Silicium.

Sn

Sn Occurs as SnO_2 'Tin Stone' Crystallized Quadratic — We have only to free it from the Gangue it is then Reduced with Carbon. It is impure from As &c — & is purified by fusing in a stream of air — the impurities oxidize & go off as AsO_3 &c.

Purifying the Tin

It approaches much in lustre + color silver - is very ductile - + stable, it is easily bent - but not like Pb - like leather - but with a grating sound (from the breaking or rubbing upon one another of crystals) - HCl - gives us action of SnCl. NO_5 - (fuming better) gives us an energetic action, + forms SnO_2. Acids Ex

Oxides

SnO = Oxide of Tin

$SnO\,SnO_2$ = Proto-Bin-Oxide

SnO_2 = Bin Oxide.

SnO_2

In two modifications - (Mono- + Penta-Basic modific).

SnO_2 obtained by precipitating $SnCl_2$ - by boiling it with H_2O, a White Precipitate.

1041

Meta Stannic acid - obtained by dissolving Tin in fuming NO_5 - - is insoluble in this acid - The two acids can be distinguished by the fact that - the first is not precipitated by NO_5 - or SO_3 but dissolves readily in these acids, While - the Meta - Stannic acid - is insoluble in these acids. the first upon heating readily converts itself into the second modification.

Ex

$SnCl_2$

$SnCl_2$ By leading Cl over Sn - is volatile + can be distilled over - can forms double salts - with Cloro bases, in which the $SnCl_2$ is the acid constituent, the salt → NH_4Cl, $SnCl_2$ is important,

Lecture 110th

These double salts are analogous to those of the Pt. metals.

$KCl, SnCl_2$ / Regular.

SnS, by precipitating. SnCl solution with H_2S, a brown black precip,

SnS_2 – by precip, $SnCl_2$ solution with H_2S – a bright yellow precip – called "Mosaic Gold" used in plating gypsum models &c.

Sn + S. SnS_2 use of

SnO.

Plays the part of a base. We have only to dissolve Sn in HCl + precipitate the solution with KOHO, by heating – it looses its H_2O. Is a black solid – forms a number of salts – on heating it does not loose its O but on the contrary if possible forms SnO_2.

SnO

1043

SnS = is a sulpho-base.

SnS_2 = is a sulph. acid

first; dissolves in NH_4S

Second; does not dissolve in NH_4S

Ex →

	HS	NH_4OHO	NH_4OCl_2	KOHO
$SnCl_2$ →	yellow	yellow insol	white. ins.	whit, sol. in ex.

Reac-tions. These are for the the Bi-Cloride - the latter for the $SnCl$.

	HS	NH_4OHO	KOHO	$HgCl_2$
$SnCl$ →	coffee brown	white. Sol. in ex.	Sol. in ex. black	white (HgCl

Ex → This brown HS precip. is soluble in NH_4S which contains NH_4SHS for it takes up yet one atom of Sulphur & forms SnS_2 which being a Sulph-Acid dissolves in NH_4S.

Charac-teristic Reaction: The characteristic and best reaction for Tin - is to dissolve the suspected subs. in HCl, on evaporation add KOHO & to this KO SnO - add - a salt of Bi_2O_3, a black Precip. is formed.

Titan.

In nature occurs as Rutil – (Anatas)(Brookit) = TiO_2 : Occur. We have to pulverize finely & reduce it mixed with C, + at a red heat – to a cloride – $TiCl_2$ → Is a yellow fluid which distills over – If we boil this solution with excess of HCl (in dilution), + continue this operation for a few hours – All the TiO_2 is precipitated as TiO_2 → to prevent traces of Fe_2O_3 TiO_2 from being precipitated – pieces of NaCl, NaOClO – are from time to time – thrown in – to prevent the Oxidation of the Iron – Metal The metal can then be reduced + forms a greyish powder – burns on heating brilliantly to TiO_2

1045

Decomposes HO at 100°

Properties + dissolves in HCl, resembles much in appearance Fe. - obtained by reducing $KFl, TiFl_2$ with metallic Potassium.

Oxides

TiO_2 = Titanic acid.

Ti_2O_3 = Sesqui Oxide of Ti.

TiO_2

TiO_2 → The greatest resemblance to ZrO_2 + SiO_2 - but is not soluble in acids (exc. HFl) (like SiO_2), can be dissolved in $KOSO_3$, $HOSO_3$ → in this way we can separate it from SiO_2 (but not from ZrO_2) → If we treat the

Separation from SiO_2 → Substance containing it with $\left.\begin{matrix}KO\\HO\end{matrix}\right\}S_2O_6$ we obtain it free from SiO_2 (but likewise with ZrO salts). The mode of separating it from ZrO_2 – when the two occur together – has not yet been discovered. ← note

It has the common property of the group → that when metallic Sn (or SnCl) is added to its solution – a compound Ti_2O_3, TiO_2 is formed of a blue color. ZrO_2 + SiO_2 do not give this reaction. The double fluorides are the most important salts – & isomorphous with Sn & Si

Eq.

Ca Fl, Ti Fl$_2$ + 4aq	Isomorphous Salts of Tic
Cu Fl, Sn Fl$_2$ + 4aq	
Cu Fl, Ti Fl$_2$ + 4aq	

Forms a compound, perhaps $NH_3 TiCl_2$? Ti + N. Forms with nitrogen a compound → Ti_3N_2?

1047

Ta, Ni.

Rare - many doubts concerning them -

They occur in Tantalit + → Columbit (Ni_2O_5). then again Euxinit, Fleurit (Norway) From these minerals it is obtained as follows! →

[margin: Ni_2O_5 + Ta_2O_5]

If we smelt the powder in $KO\,SO_3$ + $HO\,SO_3$ - the bases are drawn out - + the Ta_2O_5 + Ni_2O_5 are left insoluble behind - insoluble in acids + in HCl.

[margin: Separation from Each Other]

We must form the double fluorides - of both acids with Kalium + allow to crystallize - the two crystallize alt differently + by repeating the operation often enough we get them separately.

By glowing with Carbon + leading Cl over them — we obtain the volatile clorides. ⟵

Vanadium.

Found very rarely as $PbO\,VaO_3$ $PbO\,AsO_5$ — the history of its discovery interesting. Occurs in England with Cu. Ores, + remains behind in the Mother-liquid? Separatio The Ores are mixed + flowed with $KO\,NO_5$ — + lead through **HS gas** — **As falls out** — then NaS_2 is added + digestion carried on — VaS_2 is formed + is dissolved in the NaS. An acid precipitates out VaS_2 — + heating in the air converts it into VaO_5 — or Vanadic acid — (pure)

1049

Oxides

VaO_2 = Vanad-Oxydul
VaO_4 = " Oxyd.
VaO_5 = Vanadic Acid

VaO_5

Isomorphous with PO_5 — in nature — & can be imitated Insoluble in HO — (ex. in traces) — Combines with

VaO_5 — bases. partic. NH_4O, VO_5 — The salts are Yellow in Color (in solution). & the Acid solutions are darker. The Pearl with Borax in Oxidation flame — Yellow in Reduction flame Green

HS black sol. in Ex. } Forms an Oxy. Chloride = $Va\left\{\begin{matrix}O_2\\Cl_3\end{matrix}\right.$

Ex SO_2 — reduces the Salts — forming VaO_2 — of beautiful green color.

Appendix.

The complete Separation of the Platinum metals from one another – (translated from Bunsen's Abhandlung über das Rhodium)

By the metallurgic process of extracting Plat. from its ores, there are obtained three products: which adapt themselves admirably to the object of obtaining those rarer metals which always accompany Pt: these products are viz. 1) The 'Ore residues' – which remain behind after extracting the mass of Plat, &c. by aqua Regia; & are rich in 'Osmium & Iridium'

& hence best adapted to the purpose of extracting these metals.

2) 'Osmiridium' - obtained by mechanical 'Washing' from the first residue :- best adapted for obtaining Ru.

The Materials

3) The 'Mother liquid Residues' left behind in the aqua Regia Solution, after reduction with Fe - the Pt having previously been separated (with KCl?) - this residue is rich in Pd & Rh & can best be used to obtain these metals.

The following researches were made with a material of the last sort, for each separation 1000 grms residue were used.

The residue contains with the exception of Os. all the Pt metals. & is particularly interesting on account of it's relatively large richness in Rhodium. ←

Claus' method (the only one previous to this) involves the loss of much valuable material. To separate the Rh. from Ir., Claus used the old method proposed by "Wollaston"; a method which depends upon the "Solubility of the Ammonium (or Kalium) double salt of the Sesqui-cloride of Rhodium" in NH_4Cl. The fact, however that the salt $KCl, IrCl_2$ is taken up in considerable quantity by a solution of NH_4Cl (or KCl) saturated with a Rhodium double salt; is sufficient to

1053

awake the gravest doubts

Imperfections of Wollaston's Method of Separa[tion]

as to whether, the metal given out to be Rhodium, & to which Berzelius & Claus ascribed the atomic Weight = 52, did not contain considerable quantities of Iridium. Bunsen found it therefore necessary to leave the beaten track & to search for a more exact method of separation.

⟶ 1) Sep. of Pt + Pd from others'

Sep. of Pt + Pd from the other Pt metals

The (complete) separation of Rh, Ir + Ru - from Pd + Pt, by digestion with Aqua Regia - will not succeed with these questionable Residues; for considerable quantities of the first metals are present in the form of Hydrated sesqui Oxides – & partly present in a finely divided state – & in consequence, dissolve up with the Pd & Pt – without taking into consideration the fact that the residue left behind by this digestion is only filtered with infinite difficulty. On the contrary – it is easy to separate Pd & Pt – almost completely – & nearly pure – from the others – if the original material is mixed with 1/2 to 1/3 its weight of NH_4Cl in a Hessian Crucible; heated to complete Volatilization of the NH_4Cl, weakly glowed – until only vapors of Fe_2Cl_3 show themselves, & then removed into a Porcellain dish – & evap–

Bunsen's Method of Sep. Pd & Pt from Others

orated to a syrupy consistence with, from 2 to 3 times its weight of raw commercial NO_5 — Through the glowing with NH_4Cl the not Pt metals will be partially converted into lower clorides, Rh, Ir, + Ru, will be rendered insoluble, + the SiO_2 present as gangue, will be converted from a gellatinous to a finely powdered mass which admits of speedy filtering. The Cloric Compounds produced by the NH_4Cl give us upon digestion with NO_5 just enough HCl to dissolve the Pt to Bi clorde — while the Metallic Cu + Fe present act in so far reducing upon the Pd (in solution in NO_5-), that it is contained in the solution, not as $PdCl_2$ but as PdCl - (not precipitable, with KCl). The mass is diluted with H_2O - filtered off, the solution saturated with KCl - & the greater part of the Pt, is separated pure as KCl $PtCl_2$, which is washed out first with KCl & later with $C_4H_6O_2$, (which last must not be added to the solution), this precipitate weighed 62 grms.

Separa. of Pt. from Pd.

The filtrate is brought into a large flask (capable of being made air tight), which however, must not be more than half filled with it; if into this flask Cl gas be lead,

1057

& the same from time to time be vigorously shaken, until no more absorption takes place, all the Pd will separate out as a cinnobar-red Precipitate of $KCl\,PdCl_2$ (impure from traces of Pt & Rh, & Ir). This weighed 157 grms. The fluid from which these precipitates were obtained is now evaporated ~~not quite~~ to dryness with HCl. Upon addition of just so much HCl as was necessary to dissolve out KCl & the other sol. salts - (by aiding the solution with rubbing with a pestil), a dirty yellow colored precipitate remained behind. This is separated by filtration - boiled with NaO,HO, & a few drops of $C_4H_6O_2$, HCl is added to dissolve up the precipitate formed - & the fluid is saturated with KCl - a precipitate of 13.5 grms. of chemically pure KCl $PtCl_2$ was obtained. (The mother liquid contained only Cu & no Pt metals). The purification of the Cinnabar red precipitate of Pd. was accom. as folls: → It was dissolved in Boiling HCl, whereby a portion of the bi-Cloride, dissolves, (with Evolution of Clorine) to PdCl. It was then Evaporated with 60 grms of Oxalic acid - & dissolved up again in KCl solution, whereupon 42 grms of KCl $PtCl_2$ remained behind chemically pure.

Sep of Pd from Pt

1089

It was washed out as before. The brown fluid, was then somewhat concentrated on the H_2O bath, + upon cooling, there separated 19 grms of bright green, well formed Crystals of KCl PdCl (with some KCl), which upon testing proved to be pure from other pt metals. The fluid poured off from these Crystals - was, then neutralized Carefully with NaOHO, which gave a very slight precipitate of Fe + Cu - which was filtered off. Upon addition of KI to the filtrate, all the Pd separates as PdI. (Avoid adding an excess of KI, by taking out upon a watch -

glass - a drop of the fluid - with a cappillary tube - as long as the precipitation is incomplete, the drop appears upon a white background, brown. When complete - it is colorless. Is K I present in excess - it appears Red). Weighed 77 grms. Tested for its purity by reducing to metallic Pd - by flowing, & dissolving in NO_5 - it must dissolve completely. The whole mass is now reduced in a stream (slow) of H gas - & the Iodine came won again as HI - at last it must be strongly heated to decompose slight traces of $Pd_2 I$ - which are formed.

Pd separated

H

HI

HO

1061.

The mother-liquid from which all the Pt + Pd has been separated, may contain some Ir + Rh & it is therefore evaporated with a little KI to dryness, whereby a mixture of RhI + IrI separates – this can be dissolved up in Aqua Regia + the 2 metals separated, as will hereafter be described, by means of $\left.\begin{matrix}NaO\\HO\end{matrix}\right\}S_2O_4$, or united with the portion from which these metals will be obtained.

Sep. of **Ru** – + of Ir + Rh.

The residue.

Sep. of Ru. + of Rh + Ir

From 1000 grms – of the original material – which remained after treating with NH_4Cl + NO_5 – weighed 400 grms It was treated as follows, to get them in proper forms

& free from impurities. The method depends upon the behaviour of ZnCl to Zn. If we smelt a piece of Zinc – it rapidly is covered with a stratum of ZnO – If to this we add a metal like Iridium, the Oxide stratum hinders, – even upon pushing the Ir beneath the surface – that it should be wet, or come into contact with the Zn; if, however, we add a few grains of NH_4Cl to it – NH_3, H, + ZnCl is formed, which last dissolves the stratum of Oxide to (ZnO, ZnCl). The Zinc beneath resembles in lustre + mova-bility, Hg; as soon as the ZnCl has dissolved as much of the ZnO as it can – the coating of Oxide appears again,

Zn + ZnCl

& is instantly removed again on addition of more NH_4Cl. The smelted Zn (strewed with NH_4Cl) has the property also, in common with Hg — of wetting other metals — & if affinity exists, of forming alloys with them. By strewing NH_4Cl upon the smelted Zn, a quiet surging is kept up as NH_3 + H are given off — Many Oxides & Clorides, among which are those of the Pt metals — are, when they come into contact with this atmosphere of reducing gases (& ZnO ZnCl), rapidly reduced & dissolved up by the Zn — By this means we can

(Quantitatively) separate all metals which are not dissolved up by zinc from those which are – + among them the Pt metals –). In making the solution the Zn (in a porcelain capsule) should be constantly rotated – the gangue remains in the (ZnO ZnCl). If the regulus, immediately upon solidifying, is taken from the capsule out of the yet fluid (ZnO ZnCl) – + washed off with Ā until the Basic Cloride is all dissolved off – the gangue can be quantitatively determined, by filtration + weighing. If the regulus is not at once taken out the Capsule will be broken – owing to the unequal expansion of the Porcelain + the metal.

1065.

For quantitative Separation 20 to 30 parts of Zn

The best proportions are — for one part of the expected Pt metals — 20 to 30 pts of Zn.

Smelting with Zn

For the extraction of the residues of our NO_3-treatment — this method is excellently adapted — by smelting only once — for two or three hours the Pt metals are all extracted —

For ordinary separation 7 times part of Zn

The mode is the following:— Smelt from 3 to 3.5 Kilogrammes of com. Zn — from time to time adding NH_4Cl, in a 2 Litre Hess. Crucible, & add the 400 grms of Residue (previously weakly glowed with some NH_4Cl) & Keep the temp. for 2 or 3 hours — just above the melting point of the alloy, by adding—whenever the mass threatens to become solid—some NH_4Cl, The contents of the mass after solidifying is divided into 3 strata:— | Treatment with Zn

The Outer one—easily separated with the blow of a hammer, contains no Pt-metals—

the Next—contains some particles of the Zn + Pt alloy—imbedded in <u>ZnO ZnCl</u>;—is porous—& not very thick—

the inner stratum—consists of a frequently beautifully crystalline Regulus. (From the second—the metallic pieces are obtained, by mechanical separation with HO & added to the regulus).

To obtain this regulus, as pure as possible—it is

1067

again smelted with 500 grms of Zinc - with addition of NH_4Cl - then granulated in H_2O, + the granules dissolved in HCl (fuming), which takes place with greatest energy, in less than an hour. (The $ZnCl_2$ can be used in the next operation). The Pt metals are obtained at the bottom in form of a black finely divided powder. It contains some impurities of Zn + (of Pb Cu &c. from the Zn.). It cannot be purified by NO_3 - or aqua regia - for part of the Pt metals will thereby also be dissolved - or, they will be so suspended in the solution that filtration is impossible — If, however, the powder is treated with HCl — singularly enough, they can be obtained perfectly pure from impur. Not only Fe + Zn — but also Pb + Cu dissolve readily + with generation of H. Explanation → Electrical currents from the positive metals (Zn Fe, Pb + Cu) to the negative Pt metals — H is given off on the latter + Cl on the former — thus dissolving them. The metals (Rh Ir &c), after complete washing weighed 65 grms. (This powder possesses the property upon gentle heating to explode weakly + with evolution of light — Thenard

Note

neither H, nor Cl, nor N, ~~nor~~ HO vapor - was given off - as these are the only substances which it is possible that the metals can take up - we must suppose - that they go into an allotropic

Note condition, by our process - & that upon heating - they return again to a normal one). The powder contains mainly Rh + Ir - (as well as traces of Pt, Pd, Pb - Cu Fe + Zn)

Treatment with BaCl It is well mixed & intimately, with about 3 or 4 times its weight of completely anhydrous BaCl, & a stream of Cl-gas lead over it, at a tolerably high temperature.

Illustrated Enigma.

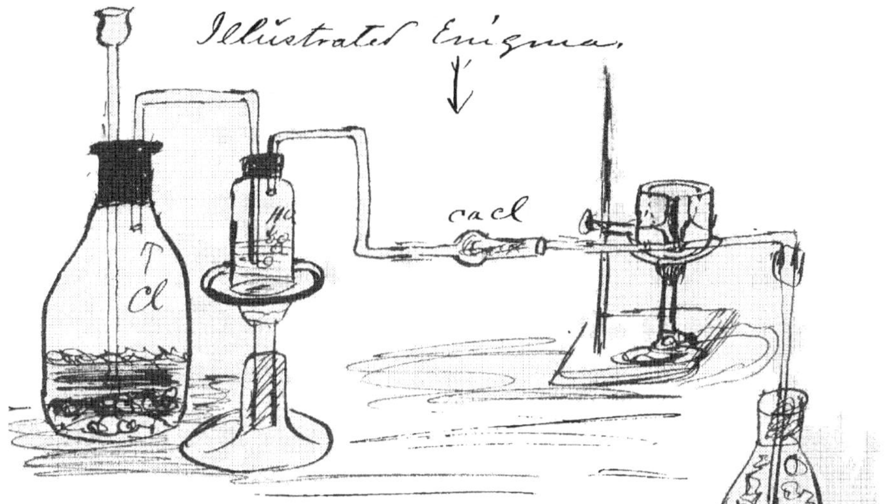

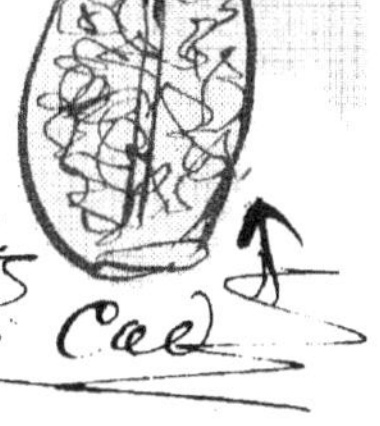

CaCl

Cl

Remove the Fe_2Cl_3 with filter paper — add little H_2O, the mass of Pt metals dissolve with evolution of Heat, readily — There remained behind 13.7 grms —, reduced (with H + then) with zinc — + heated with HCl, there remained behind 4.5 grms. — Ruthenium. from the original 65 grms — in three hours + 4 ordinary burners — 57 grms Pt metals were dissolved out (+ 4 13 grms.

Ru separated

of 85% MnO_2 were consumed.
From this solution all BaO
is removed by careful addi-
Purified tion of SO_3 - (see addition
from of KI). The Pt metals are
Fe, Cu, &c now freed from all the others
by reduc. by reducing with H gas - is
with H - completed in 5 or 6 days -
of 100 grms Pt metals are
present - (temperature nearly
100°C - during the operation
the constant HCl bath used).
Pt + Pd chiefly separate
first, Rh comes next -
& the last portions contain
mostly Ir. It is best to
break off the operation, when
the fluid has taken on a
greenish yellow color - &
add the last portions of Ir (after
drawing them out ~~[struck out]~~ by

evaporating the contents of the flask to dryness, glowing with $NaOCO_2$ + treating the solution with aqua Regia) to the portion afterwards to be opened by BaCl. The reduction is hastened by removing the HCl formed, from time to time – by concentrating the fluid – (have a care against Explosions). The separated Pt metals – are treated with aqua Regia – Pt + Pd thus dissolved out – + separated – as above given. The traces of Rh + Ir in the mother liquid which are precip. by continued boiling with KI. as Iodidides – these are dissolved in Aqua Regia + added to the part insoluble. The insoluble + Partly oxidized Pt metals — are again reduced in H gas — + treated as before described with BaCl. Boil the solution — freed from BaO — with NaO, HO; + separate the last traces of Pd + Pt. + there remains only <u>Rh + Ir</u> to be separated.

The brown-red fluid is for this purpose — evaporated with HCl, filtered — + to it, is added a <u>great excess</u> of $\frac{NaO}{HO} S_2O_4$, and the whole left standing <u>in the cold</u>, several days. The double salt of Rh (i.e. $NaOSO_2, RhOSO_2$) separates slowly — in the form of a lemon yellow precip. The solution becomes lighter & lighter - & finally almost colorless; as its color changes, so also does that of the Precip. become lighter - This precip. contains, after being well washed - the Rh - almost entirely pure. { Rh pure

Upon heating the fluid - gently - a yellow-white precip. separates - which consists mainly of Rh but contains some Ir. After filtering off this precip. if the solution is evaporated to a small volume - on the HO bath - 2 precip's separate -

1st a Kurdy slowly separating yellowish-white Precip - nearly chemically pure Ir - only

Ir. Separa. | containing the slightest traces of Rh

2nd. A heavy crystalline powder, quickly separating. It can readily be separated from the other by decantation, + weighed 16 grms. It showed all the reactions of Ir. but besides this other peculiarities – so that

Note | Bunsen in it – suspects a <u>new metal</u>

Without counting these 16 grms – the precipitate (1st one) weighed <u>99.5 grms</u>. The mother liquid is free from Pt metals.

The complete separation of Rh + Ir. from each other is accomplished by treating the yellow precipitates with concentrated SO_3. They are brought in small portions, in the SO_3 - heated in a porcellain capsule, until the SO_2 is gone off. + evaporate until all the free SO_3 is gone, (upon the Sand bath), + until (NaO HO S_2O_6) is formed. Upon boiling the mass with HO, all the Ir - dissolves up as sulphate with a chrome-green color - while the Rh - remains behind as flesh- red double salt of Soda + RhO. The latter must be washed out with cooking with Aqua Regia + decanted with HO. Is insoluble in HO, HCl, NO_5 + Aqua Regia.

1077

These two substances Rh + Ir are now completely separated — The Rhodium salt weighed 33.2 grms.

The following is the weight of the various precips from 1000 grms original material.

KCl PtCl$_2$	=	117.5 grm
Pd. I	=	77.0 "
KCl; PdCl	=	19.0 "
NaO SO$_3$ RhO SO$_3$	=	33.2 "
Ir$_2$O$_3$	=	9.4 "
Ru (+ Ir)	=	11.5 "

The first yellow precipitate obtained in the cold {by $\frac{NaO}{HO}$} S_2O_4 gave, by this treatment, the Rh, quite pure. The second + third precipitates — containing much Ir

give a very pure Rh - but still with traces of Ir.
The products therefore obtained by this treatment with SO_3 (which betray their impurity of Ir by their somewhat brownish color) are collected by themselves - the Rhodium is separated therefrom by glowing - & the metal is treated again with BaCl. (& repeat the operation of separation again).
The green solution containing only Iridium is gradually heated, first over a free burner, in a porcellain caps - & afterwards - upon the sand-bath, to remove the excess of SO_3, & finally - the crucible & its contents are heated highly in a hessian crucible -

whereby there is formed – $NaOSO_3$ + Ir_2O_3 – Upon boiling the mass with HO – the latter remains behind as a black insoluble powder – & can easily be washed out by decantation; it weighed 9.1 grms. Results are found on page 1077.

In these separations – the new suction pump. & the Continuous HO Bath materially shortened the labor.

Winter Semester /69

Table of Contents. 1880

1081

Page	Lec. 11th	Page	Lecture 16th
64-69	Sp. Gr. of gases.	94-100	Catalysis, Solution Crystallization & Distillation Dyalysis.
70-73	Lec. 12th. Ditto. Sp. grav. of Vapors.		Lecture 17th
74-81	Lec. 13th. Combinations - various Kinds of. Mechanical Mixtures &c	101-107	Dialysis. Chemical Compds. Chemical affinity, & influence of Phys. agents on.
82-85	Lecture 14th. Diffusion Tension of Vapos & Gases	108-111	Lec. 18th. Electricity - connection with Chem. action - Contact action - analysis - Kinds of.
86-93	Lec. 15th. Properties of gases. absorption Condensat. How to collect gases for Analysis Etc.	112-117	Lecture 19th Oxygen. Phlogistan Theory. Compounds etc

1083

Lec. 38th	page	Br & Oc.	page
Cl. Compds		BrO_5 - - -	
HCl.		Iodine	
Props &		Manufac -	
Manufac.	350-356	& Props.	388-397
39th		43rd	
HCl cont.		HI,	
Cl & O comp.		analysis	
ClO. - - - -	357-365	of Iodides with Cl.	
40th		I & Oxygen	
$HOClO_5$ —		IO_5 - - - -	398-407
$HOClO_7$		44th	
ClO_3 - - - -	366-375	IO	
41st		IO_7, HO	
ClO_4 -		IO_4	
Bromine		ICl, ICl_3	
Manufac &		ICl_5 —	
Properties		Fluorine	
Compds.		HFl. …	408-419
HBr. …	376-387	45th	
42nd.		HFl —	
analysis of		Uses.	
Br. Compds		Sulphur	
		Props & Man.	420-426

46th	page	52nd	page
Sulphur.		Selenium	
HS.	429-439	SeO_2 + SeO_3	
47th		Tellurium	480-4
HS: HS_2		53rd	
S + Oxygen		TeO_2 + TeO_3	
SO_2	440-448	Nitrogen	
48th		Props & Manufac.	
SO_2 - Analyses of with I. Solution	449-456	NH_3	490-499
49th		54th	
I. Titration Contin.		NH_3	
SO_3 -		NH_4 + Compds..	500-508
50th		55th	
SO_3	457-470	NCl_3 ...	509-514
51st		56th	
S_2O_2		NO_5 - HO	
HO, S_2O_5 —		Props &	
HO S_3O_5 —		Manuf.	575
HO S_4O_5 —		Salts &c	524
HO S_5O_5			
S + Cl			
S_2Cl + SCl	472-47		

1087

64th	page	CO_2	page
$AsCl_3$, AsS_3, AsS_2 Sb	594–605	Properties & Manufacture	632–641
65th		**69th**	
SbH_3 SbO_3, SbO_5 SbO_4, Sb Poisoning	606–613	CO_2 Liquidification Solidification Metamorphism of Rocks &c.	642–648
66th		**70th**	
Mirror of Sb. Separation of As from Sb. $SbCl_3$ & $SbCl_5$ SbS_3 & SbS_5 Carbon	614–624	Analyses of CO_2 – mode of the Atmosphere Analyses of its constituents	649–656
67th		**71st**	
Modifications of C. Graphite – Organic C. & Diamond	625–632	Atmosphere Constitution of in different places	657–663
68th			
Charcoal			

1089

79th	page
HbCy.	
$Cy_3O_3, 3HO$.	
Urea –	
C_2N, S = Pharoah's serpent	718–725
80th	
Boron	
Occurrence	
Props. etc.	
BoO_3 Manufacture	726–731
81st	
Tests for BoO_3	
Uses, Soldering	
H, Bo + Cl –	
Fl –	
Method of Organic analyses, Silicium. SiH_2	732–741
82nd	
SiO_2. Occurrences.	

Modification kinds of silicates.	page
Fl_2Si, Zirconium	742–751
83rd	
Zr. ZrO_2	
Metals –	
their grouping = general properties.	
Na. Occurrence, manuf. NaO + NaO_2 + salts of NaO	752–763
84th	
Salts of NaO.	
NaCl, Occurrence.	
General tests for metallic oxides –	
Reactions for Na . . .	763–771

85th	page	88th	page
Kalium, Occurrence manuf. & properties. KO. + Salts Gunpowder	771–781.	Ca, Ba + Sr Group. Ca. CaO– uses of – CaOCO2 – Other Salts.	802–811
86th		89th	
Kalium. Reactions. Spectrum & Flame. Caesium & Rubidium. Mg from p....	782–792	Cement. Glass– diff. kinds of. Reactions for CaO + Strontium– Occurrence &c. SrO + Salts–	
87th			
Lithium. LiO + Salts– Separation from alkalies Reactions. Mg. MgO + Salts – –	793–801	Barium. BaO + Salts Tests– Spectra &c. Separation from Each other in the wet way– by Cy Ho2 –	812–823

1091

90.th	page	92.nd	page
Ba O Salts. Ba Halloids. Electrolysis, mode of separating Na. K. Rb. Cs. Mg. Li, Ca. Sr. Cr & other metals by the battery	824–831	Spectrum analysis of the Sun, Stars, & Nebulae — Alkilimetry Acidimetry Er & Y } Ce. La. Di } Groups	846–853
91.st		93.rd	
Light in its relation to Chemistry — Laws of absorption & Emission — Spectrum of Sol. liq. & Gases — Spectroscope & Spectrum analysis.	832–843	Separation from other groups & from each other of Er & Yt. Er. Yt. Group {Ce. La. Di Separation from other groups & from each other. Ce CeO & Salts — La & Di — Salts &c	854–860

www.ingramcontent.com/pod-product-compliance
Lightning Source LLC
Chambersburg PA
CBHW020907210326
41598CB00018B/1796